1

Physics of Thin Films

Advances in Research and Development

VOLUME 13

Physics of Thin Films

Advances in Research and Development

Edited by

MAURICE H. FRANCOMBE

Research and Development Center
Westinghouse Electric Corporation
Pittsburgh, Pennsylvania

JOHN L. VOSSEN

John Vossen Associates
Technical and Scientific Consulting
Bridgewater, New Jersey

VOLUME 13

ACADEMIC PRESS, INC.
Harcourt Brace Jovanovich, Publishers
San Diego New York Berkeley Boston
London Sydney Tokyo Toronto

ACADEMIC PRESS, INC.
1250 Sixth Avenue, San Diego, California 92101

United Kingdom Edition published by
ACADEMIC PRESS INC. (LONDON) LTD.
24–28 Oval Road, London NW1 7DX

LIBRARY OF CONGRESS CATALOG CARD NUMBER: 63-16561

ISBN 0–12–533013–8 (alk. paper)

PRINTED IN THE UNITED STATES OF AMERICA

87 88 89 90 9 8 7 6 5 4 3 2 1

Contents

Ionized Cluster Beam Deposition and Epitaxy

Toshinori Takagi

The Activated Reactive Evaporation Process

R. F. Bunshah and C. Deshpandey

Ion-Beam Processing of Optical Thin Films

Ursula J. Gibson

Contributors to Volume 13

Numbers in parentheses indicate the pages on which the authors' contributions begin.

CAROL I. H. ASHBY (*151*), Sandia National Laboratories, P.O. Box 5800, Albuquerque, New Mexico 87185

N. BRASLAU (*199*), IBM Thomas J. Watson Research Center, P.O. Box 218, Yorktown Heights, New York 10598

R. F. BUNSHAH (*59*), Department of Materials Science and Engineering, University of California at Los Angeles, Los Angeles, California 90024

C. DESHPANDEY (*59*), Department of Materials Science and Engineering, University of California at Los Angeles, Los Angeles, California 90024

J. L. FREEOUF (*199*), IBM Thomas J. Watson Research Center, P.O. Box 218, Yorktown Heights, New York 10598

URSULA J. GIBSON (*109*), Optical Sciences Center, University of Arizona, Tucson, Arizona 85721

TOSHINORI TAKAGI (*1*), Ion Beam Engineering Experimental Laboratory, Kyoto University, Sakyo, Kyoto 606, Japan

J. M. WOODALL (*199*), IBM Thomas J. Watson Research Center, P.O. Box 218, Yorktown Heights, New York 10598

Preface

This volume of *Physics of Thin Films* contains five articles, four of which deal primarily with the influence of ions or of optical energy on the deposition, properties, or etching of thin films. In the first article T. Takagi describes recent developments in the field of ionized cluster beam deposition, a growth method pioneered by his group in Kyoto. After outlining the experimental features and the physics of the process, extensive examples are discussed of its application to films of metals, insulators, and semiconductors. The novel properties of ionized clusters in enhancing adhesion, film orientation, and crystallinity and in lowering temperatures for epitaxy are reviewed.

The second article by R. F. Bunshah and C. Deshpandey addresses a closely related deposition process, viz., activated reactive evaporation. The basis of this technique involves evaporation of reactive elements or subcompounds through an activated gas plasma zone, which is sustained by means of an electron beam. The process is highly successful for growth of stoichiometric refractory oxide, carbide, and nitride films and yields adherent coatings displaying excellent mechanical properties.

In recent years there has been rapidly increasing interest in the important role of ions, not only in growth processes, but also in influencing film properties. In the third article U. J. Gibson considers in general the key consequences of ion bombardment on the characteristics of thin films and discusses in particular the effects of controlled ion bombardment during growth on the mechanical, chemical, microstructural, and optical properties of films used in optical components. She shows that many of the critical properties, especially of dielectric films, such as density, adhesion, stoichiometry, refractive index, and optical absorption, can be varied systematically and optimized in ion-assisted deposition processes.

The fourth article, by C. I. H. Ashby, reviews the status of the field of laser-induced etching of thin films. Generation of high-resolution patterns in films by means of ion or laser etching is of considerable importance, especially for fabrication of high-speed integrated circuits. Her article presents first a comprehensive discussion of the mechanisms of laser-induced etching, proceeds next to factors which influence resolution, rates, and selectivity, and finally details recent applications of the method to films of polymers, metals, semiconductors, and other inorganic materials.

The fifth and final article, by J. M. Woodall, N. Braslau, and J. L. Freeouf, deals with the important technological problem of contacts to GaAs devices. The authors discuss the limitations of present models in accounting for the properties of metal contacts and propose a new effective work function model which suggests that observed Schottky-barrier heights are determined by the work function of surface phases, generated as products of contamination or of contact processing, which form an interface with the semiconductor. Problems of obtaining low ohmic tunneling resistivities with current alloy contact technology are reviewed, and a new ohmic contact structure for GaAs is described in which a graded bandgap layer of GaInAs is epitaxially grown on GaAs by MBE.

M. H. Francombe
J. L. Vossen

Editors' Note

We wish to note the retirement of George Hass from the position of Senior Editor of this series. Dr. Hass was the founder of Physics of Thin Films and edited the first volume in 1963. Since then he has guided the technical direction of the series and has contributed numerous articles in the field of optical films. We look forward to his continued advice and participation in the future.

Ionized Cluster Beam Deposition and Epitaxy

TOSHINORI TAKAGI

Ion Beam Engineering Experimental Laboratory
Kyoto University
Sakyo, Kyoto 606, Japan

I. Introduction

The ionized cluster beam (ICB) deposition and epitaxial method, which has been developed by Takagi *et al.* (*1*), is an ion-assisted technique by which high-quality films of metals, dielectrics, active semiconductor materials, and some organic materials can be formed at a low substrate temperature in a technical-grade vacuum system.

In ion-assisted film formation, ions transfer energy and charge to a substrate and a depositing film surface. The role of ions becomes of pri-

TABLE I

INFLUENCE OF IONS ON FILM FORMATION

a. Kinetic energy is converted to
 Sputtering energy
 Thermal energy
 Implantation energy
 Migration energy on substrate surfaces
 Creation energy of activated centers for nucleus formation

b. The presence of ions has a great influence on
 Critical parameters in the condensation process of the film formation such as nucleation, coalescence, etc.
 Chemical reaction, even without additional acceleration voltage, and even when only a few percent of ionized particles are included in the total flux.

mary importance for film formation (*2, 3*), and it may be described in terms of kinetic energy and ionic charge, as listed in Table I. The optimum value of the kinetic energy is different for particular combinations of deposits and substrate materials and for applications purposes. It is in the range of a few to a few hundred electron volts, as listed in Table II. The

TABLE II

OPTIMUM CONDITIONS FOR THE KINETIC ENERGY OF IONS INCIDENT ON THE SUBSTRATE FOR FILM FORMATION

Conditions	Required incident ion energy	Result
Deposition	Less than the energy corresponding to the sputtering rate $S(E) = 1$ Larger than the energy at which the sticking probability becomes too low	Optimum value of kinetic energy: a few to a few hundred electron volts
Surface cleaning	Larger than the energy of adsorption on the substrate surface, i.e., 0.1–0.5 eV for physically adsorbed gases and 1–10 eV for chemically adsorbed gases	
Good-quality film formation	In a range where enhanced adatom migration influences properties of the deposited film; suitable ion bombardment affects the growth of nuclei; a suitable number of defects or atomic displacements near the substrate surface contribute to film formation during the initial stage	

kinetic energies of the source particles are converted to sputtering energy, thermal energy, implantation energy, migration energy, and energy for creating activated centers for nucleus formation. It is reported in the case of vacuum evaporation that the bombardment of ions on the depositing surface greatly influences the critical parameters of the condensation process, and consequently causes changes in film characteristics such as physical, optical, magnetic, or crystallographic properties (*4, 5*). The presence of ions has the effect of enhancement of the film formation activity and the chemical reaction activity of the evaporated materials. It plays an important role at the initial stage of film formation, which is especially useful to the formation of good-quality thin films of oxide, nitride, carbide, etc. (*3*). These effects are predominant even when only a few percent of the ionized particles are included in the total flux (*5*).

In the ICB technique, macroaggregate atoms (clusters) of deposit material vapor are utilized instead of atomic or molecular state particles (*6, 7*). The vaporized-material clusters containing 500–2000 atoms loosely coupled together can be formed from supercondensation phenomena following adiabatic expansion of the vapor through a cylindrical nozzle. The characteristics of the cluster are different from those of a droplet (liquid particle), because the droplet contains 5×10^8–5×10^9 atoms and they are closely coupled to each other (*8*). The clusters are partially ionized to the singly charged state by electron bombardment, and the ionized clusters are accelerated toward the substrate by a high negative potential. Neutral clusters also drift toward the substrate at the ejection velocity. When the clusters bombard the substrate, they break up into atoms. Each of the atoms of the clusters has an average energy of $\bar{E} = e\,V_a/N$, where e is the electric charge, V_a is the acceleration voltage, N is the cluster size (number of atoms per cluster). By controlling V_a it is possible to provide each atom with a suitable energy for film formation. The beams have an extremely small charge-to-mass ratio, and they are well collimated at high intensity. These features allow the cluster beams to transport a large mass at very low energy without any problems caused by the space-charge repulsion force.

In film formation by the ionized cluster beams, the enhanced adatom migration effect (one of the effects listed in Table I) can be achieved effectively by controlling the acceleration voltage. The charge on the cluster is sufficient to influence film formation, although the number of electric charges in the total beam is very small. By using these effects, the ICB technique could be applied to prepare many kinds of active and passive thin-film devices.

In this article the basic operating mechanism of the ICB equipment is described. The ICB film-formation mechanisms are also explained by describing the film characteristics that can be achieved.

II. Equipment for Ionized Cluster Beam Deposition and Epitaxy

1. ICB Deposition System

A typical schematic diagram of the ICB equipment is shown in Fig. 1. The molecular beam source utilizes molecular flow by using a Knudsen cell, but the cluster beam source utilizes viscous flow by using an ejection source. The dimensions and factors for design of the cluster source are summarized in Fig. 2. The clusters grow by collisions with surrounding vaporized atoms in the nozzle region. The nozzle diameter, D, has to be larger than the mean free path, λ, of the atoms in the crucible. The ratio of the inner pressure P_0 of the crucible to the vapor pressure P outside the crucible must be larger than 10^4–10^5. Since P decreases at least to the background pressure in a vacuum chamber, that is, 10^{-7}–10^{-5} torr, the inner pressure, P_0, has to be kept at 10^{-2}–several torr. It is also desirable to fix the nozzle thickness, L, as $L/D = 0.5$–2.0 in order to keep the pressure ratio P_0/P at a high value and to achieve enough collisions of the vaporized atoms in the nozzle region. A simple nozzle shape, a cylindrical type with a diameter $D = 0.1$–2.0 mm, is enough to form a cluster beam

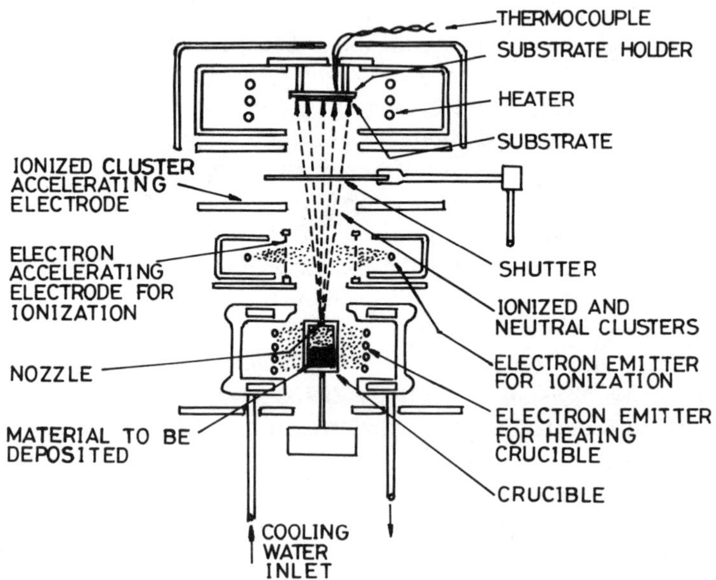

FIG. 1. Schematic diagram of the ICB equipment for deposition and epitaxy.

with a high drift velocity (*9*). A multiple-nozzle system, where each nozzle satisfies the dimensions mentioned above, can be used for film deposition over a large area with good uniformity. The range of the source temperature is adjusted in order to produce a vapor pressure P_0 of the order of 10^{-2}–several torr. For heating the crucible, direct resistive heating, electron bombardment heating, or hybrid methods can be chosen according to the application.

The ionized clusters are produced by electron impact in an ionization electrode system located above the crucible. The ratio of ionized clusters to the total clusters can be adjusted by changing the electron current (I_e) for ionization. For example, as shown in Fig. 3, the degree of ionization obtained for a single nozzle by assuming that only one atom in a cluster is ionized is 5–7% at I_e = 100 mA, 7–15% at I_e = 150 mA, and 30–35% at I_e = 300 mA (*10*). The ionized clusters are accelerated by the acceleration

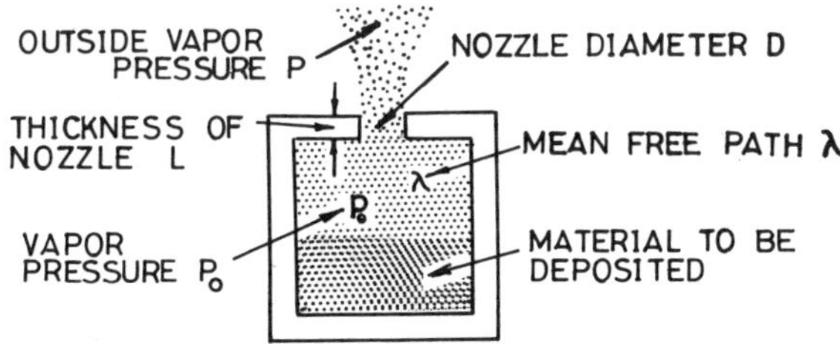

Cluster beam

Supersaturated vapor in an adiabatic expansion

$D \gg \lambda$

(for example: D = 0.1 - 2 mm)

$\frac{P_0}{P}$ > about 10^4—10^5

(for example: $P = 10^{-7}$—10^{-5} Torr, $P_0 = 10^{-2}$ — several Torr)

experimentally

$\frac{L}{D} = 0.5 — 2.0$

FIG. 2. Dimensions and factors for designing the cluster beam source.

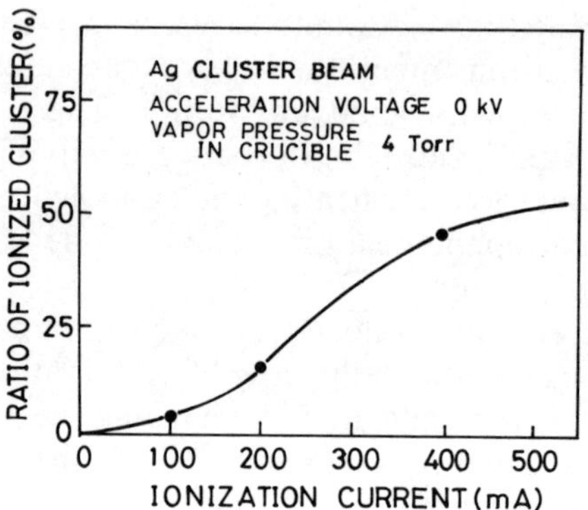

FIG. 3. Relation between the electron current for ionization and the ionization ratio.

voltage (V_a) to the substrate. The ionized clusters and the neutral clusters that are not ionized in the ionization electrode system bombard the substrate. The ionized clusters have a kinetic energy corresponding to the acceleration voltage, whereas the neutral clusters have a kinetic energy corresponding to the ejection velocity.

In reactive ICB (RICB) deposition for hydride, oxide, nitride, or carbide film formation, the deposition is carried out in a low-pressure reactive gas in the range of 10^{-5}–10^{-4} torr. In this pressure range a plasma is not produced in the chamber. If a plasma occurs in the chamber, the clusters are destroyed by collision with energetic particles, and the advantages of the ICB technique are lost.

A view of an ICB experimental deposition system, which is commercially available, is shown in Fig. 4 (*11*). The system has four crucibles which can be moved sequentially into the ion source by remote control. Deposition conditions such as crucible temperature, deposition rate, acceleration voltage, electron current for ionization, substrate temperature, film thickness, etc., can be controlled by a computer system. A dual cluster-beam system having two groups of crucibles also has been developed. A single crucible system with multiple nozzles which form a ribbon beam with high uniformity has been developed to deposit films on large substrates, as shown in Fig. 5. A similar idea was applied to an industrial production system (*12*).

2. Cluster-Beam Formation

With regard to cluster formation, the source material in the crucible is heated to a high temperature. When vaporized material is ejected through

FIG. 4. ICB experimental deposition system.

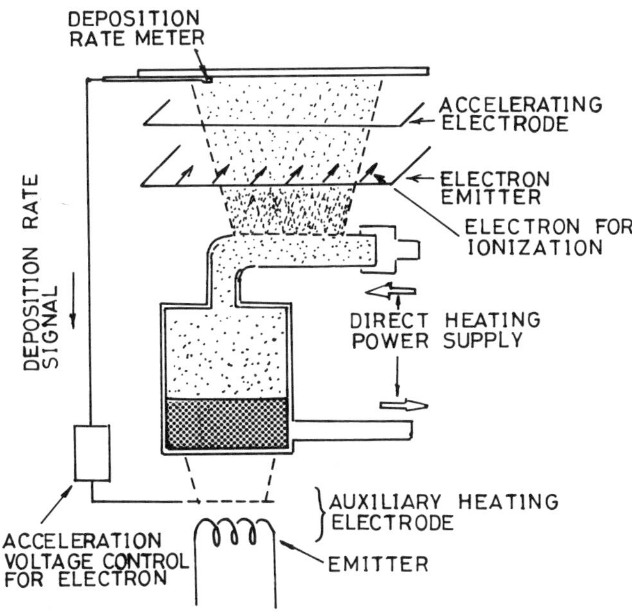

FIG. 5. Schematic diagram of the ICB deposition system using a crucible with multiple nozzles.

a nozzle into a high-vacuum region, the vapor atoms collide and transfer their energies to each other, and they are in a supersaturated state (*13, 14*). The atoms, which lose their energies by the collision during the adiabatic expansion, start to aggregate to form nuclei. The nuclei smaller than a critical nucleus size are not stable and they break up into small pieces. However, the nuclei formed in the supersaturated state that are larger than the critical size grow to form clusters (*15*). The growth of the clusters occurs near the nozzle and then slows down after reaching the maximum rate, and finally ceases in the region where the pressure decreases and collision frequency becomes lower. In a high-vacuum region most of the clusters drifting to the substrate keep their sizes constant because there is little collision between clusters and residual gas atoms. The clusters thus formed have high kinetic energy because (1) the thermal energy of the expanding vapor is converted into kinetic energy and (2) the vapor aggregates into a cluster (*16*).

The cluster formation mentioned above is based on a process which is related to adiabatic expansion and homogeneous nucleation. According to the classical theory of droplets, a vaporized-material cluster consisting of several hundreds to several thousands of atoms would be difficult to form, assuming that the surface tension of the bulk liquid applies. However, a recent theory predicts that the surface tension tends to decrease with a decrease of the size of the droplet (*17*) and that the surface layer of the cluster cannot be clearly defined (*18*). Therefore, it should be noted that the formation of the clusters containing fewer than several thousand atoms is different from that of droplets, since the surface tension of the bulk liquid may not apply for such clusters.

3. Size of the Clusters

The cluster size, i.e., the number of constituent atoms in a cluster, can be estimated by measuring the kinetic energy and the velocity of the cluster. The kinetic energy E_{kin} of a cluster consisting of N atoms can be written as

$$E_{kin} = \frac{1}{2} NmU^2 \tag{1}$$

where m is the mass of the atom and U is the flow velocity (the ejection velocity). Since E_{kin} and U can be measured experimentally, the cluster size can be calculated by using Eq. (1). The cluster size N can also be estimated roughly from E_{kin} divided by the thermal energy E_{th} of the constituent atoms of the order of kT_0. In this rough estimate, the kinetic energy of the clusters consisting of N atoms is given by

$$E_{\mathrm{kin}} = NE_{\mathrm{th}} \tag{2}$$

The energy of the cluster can be measured by an electrostatic 127° energy analyzer and retarding field methods. In the measurement of the kinetic energy, the ionized cluster must be accelerated to an energy E by the accleration voltage V_a. Therefore, the kinetic energy of the cluster (E_{kin}) becomes

$$E_{\mathrm{kin}} = E - eV_a \tag{3}$$

where e is the electric charge.

Figure 6 shows the energy distribution of the ionized clusters measured by a 127° energy analyzer as a function of the metal vapor pressure P_0 in the crucible (*19*). Two kinds of peaks are observed in the spectra: one has an energy close to zero and the other has an energy of 80–170 eV. The intensity of the higher-energy peak, which corresponds to the clusters, increases beyond the critical pressure and then decreases with a further increase in the pressure of metal vapor in the crucible. In our experiment the nozzle diameter D was 1 mm and $L/D = 1$. The clusters begin to be observed at a vapor pressure of about 0.5 torr, and the intensity of the cluster beam increases to a maximum at $P_0 = 4.7$ torr. At pressures higher than 5 torr the intensity of the beam begins to decrease. This is

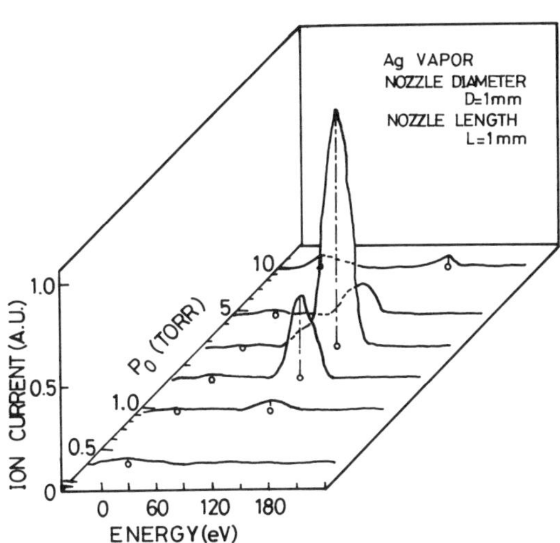

FIG. 6. Energy distribution of the ionized cluster beam as a function of metal vapor pressures in the crucible.

considered to be because the difference in pressure inside and outside the crucible decreases owing to the high rate of ejection from the crucible. The maximum and minimum values of P_0 observed can be extended by changing the shape or the dimensions of the nozzle.

Figure 7 shows the influence of the nozzle diameter on the energy and intensity of the clusters. The mean free path of the atoms in the vapor at an inner pressure of 1.25 torr is of the order of 0.01 mm, and therefore the diameters of all the nozzles tested are sufficiently large for the occurrence of the many collisions necessary for cluster growth during the expansion. In the case of the 2 mm diameter nozzle, clusters were formed most effectively. However, no peak at higher energy is observed for a crucible with a diameter larger than 5 mm, including an open crucible, because the pressure difference between the inside and the outside of the crucible is insufficient. Under these conditions the ejected vapor consists of atoms, molecules, and/or small clusters comprising several atoms instead of larger clusters.

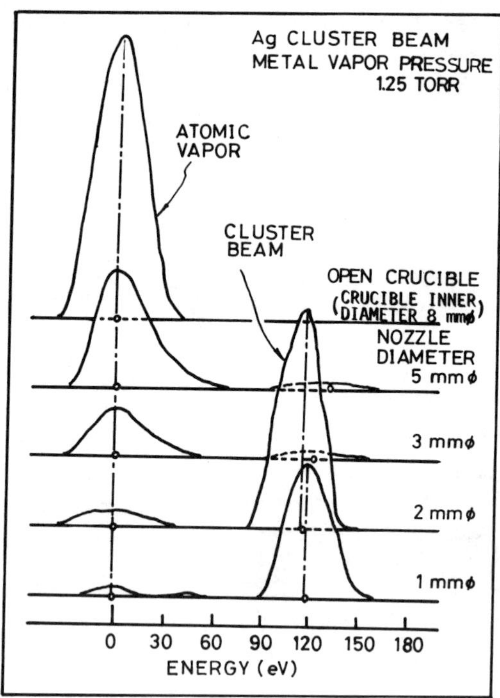

FIG. 7. Dependence of the energy and the intensity of the ionized cluster beam on the nozzle diameter.

By using the retarding field method, the energy of the Ag cluster ions was measured, as shown in Fig. 8. The distribution of the beam energy was calculated by differentiating the ion current with respect to the retarding potential. The energy of the cluster is found to be in the range of 40–170 eV (*10*).

For the measurement of the velocity U, a rotating disk method was used. The velocity was calculated by measuring the shifted distance of the deposits on the disk. Figure 9 shows the experimental results between the inner pressure P_0 and the most probable velocity of the clusters. As shown in the figure, as P_0 increases, the velocity U approaches the value $U_{\max} = \sqrt{5kT_0/m}$, where T_0 is the source temperature and m is the mass of a constituent atom in the cluster.

By using experimentally obtained values of E_{kin} and U, the cluster size was calculated to be 500–2000 atoms/cluster. By using E_{kin} measured by the retarding electric field method and the thermal energy E_{th} of the constituent atoms of the order of 0.1 eV, the size of the cluster is calculated to be of a similar value to that obtained by the above-mentioned methods. In the case of CdTe, a similar result was also obtained (*20*). By using the time-of-flight method and the negative-pulse applied-voltage method, the cluster size of Pb was found to be 500–1000 (*6*).

The cluster size of Ag can also be measured by observing the individual clusters deposited on an electron microscope mesh. The cluster was collected on a carbon film on a copper grid mesh, which was cooled to a

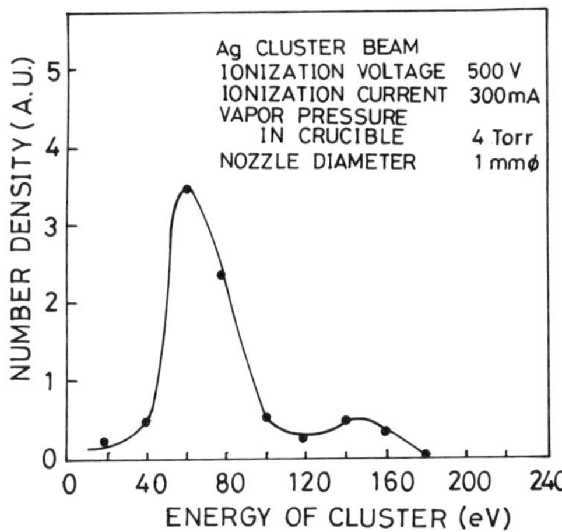

FIG. 8. Energy distribution of the ionized Ag cluster beam measured by the retarding field method.

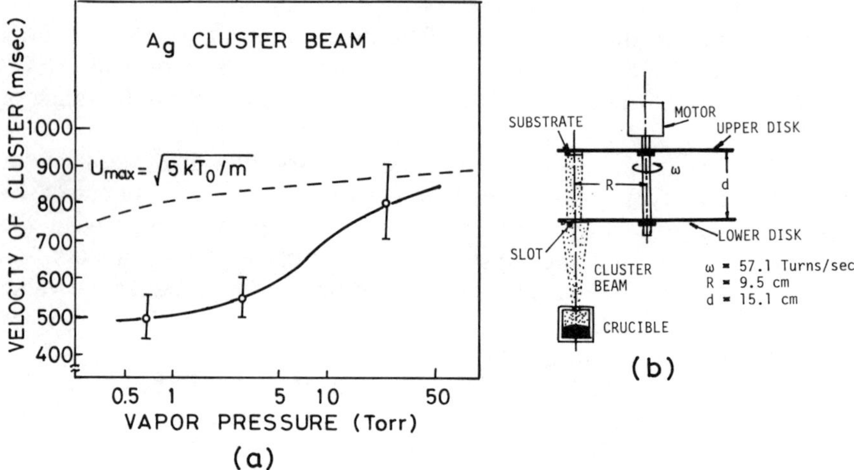

FIG. 9. (a) Relation between the inner pressure of the crucible and the most probable velocity of the clusters, and (b) the rotating disk method.

liquid nitrogen temperature. The deposition time was determined so as to prevent the coalescence between collected clusters. Figure 10 shows histograms of the cluster diameter distribution as a function of the vapor pressure in the crucible, P_0. In the range of P_0 below 3 torr, the mean value of the cluster diameter is about 50 Å. On the assumption that the cluster deposited on the mesh is a hemisphere and the deposited cluster has the same structure and lattice constant as those of bulk silver, the cluster size is calculated to be about 2000 atoms. Taking account of the geometrical change of the clusters collected on the mesh, the difference of the density of the cluster from that of bulk and so on, the actual cluster size is considered to be smaller than 2000. The observed cluster diameters increase gradually as P_0 increases. At $P_0 = 6$ torr, a wide spread of the distribution of the cluster diameters can be seen. It is likely that the condition or mechanism of cluster formation undergoes some changes in this pressure region.

4. Structure of the Clusters

In each cluster, a considerable portion of the constituent atoms are located on its surface. Consequently, the physical properties of such vaporized-material clusters are quite different from those of liquid-state droplets and bulk materials. It may be said, therefore, that clusters might be a new phase of material, that is, the fifth state of matter (*21*).

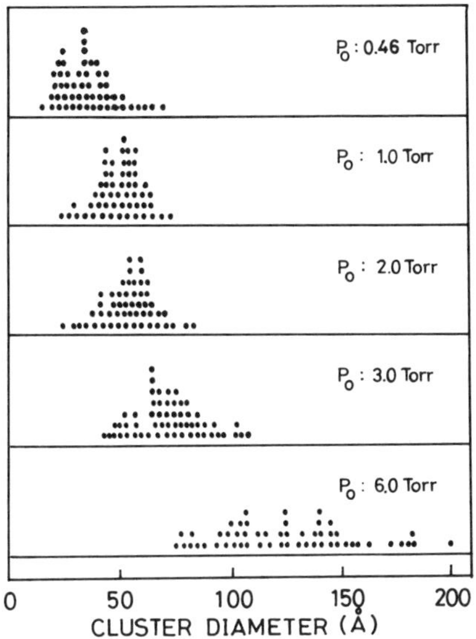

FIG. 10. Histograms of the cluster diameter distributions obtained by TEM observations.

The phase of clusters can be determined by analyzing the electron diffraction patterns of cluster beams (*22*). The cluster source was mounted in an electron microscope, and the cluster beams were formed by ejecting the vapor through a cylindrical nozzle 1 mm in diameter and 1 mm in length. The background pressure outside the crucible was reduced low enough to ensure supersonic expansion. Sb was used as a test material. The Sb cluster beam was collimated with the aperture and crossed by a 45 kV electron beam. The microdensitometer trace of the obtained diffraction pattern is shown in Fig. 11. The pattern is a broad halo, and no Debye rings indicating crystalline structure could be seen. This indicates that the vaporized Sb clusters formed in this way are amorphous. A pattern from a polycrystalline Sb film prepared by vacuum evaporation also is shown in the figure. The abscissa indicates the scattering parameter given by

$$s = 4\pi \sin(\theta/2)/\lambda \tag{4}$$

where λ is the wavelength of the electron beam and θ is the scattering angle. The diffraction pattern was Fourier transformed and a radial distribution function was calculated to obtain the interatomic distance. Figure 12 shows the comparison of the interatomic distance of the Sb clusters together with those obtained for the amorphous and the crystalline films.

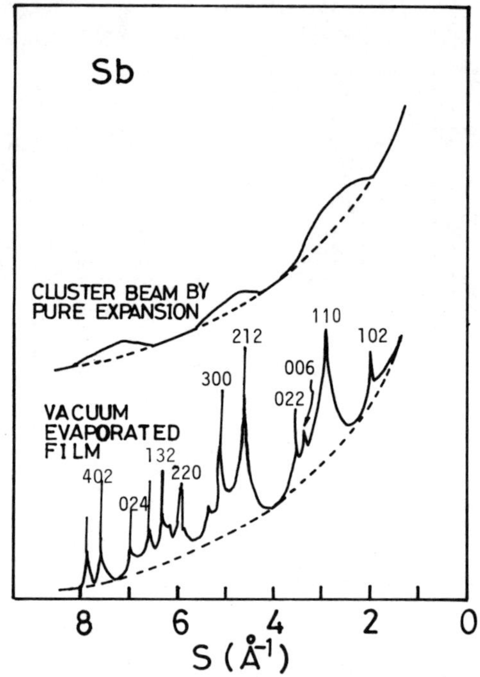

FIG. 11. Diffraction patterns of Sb films deposited by cluster beams and vacuum evaporation.

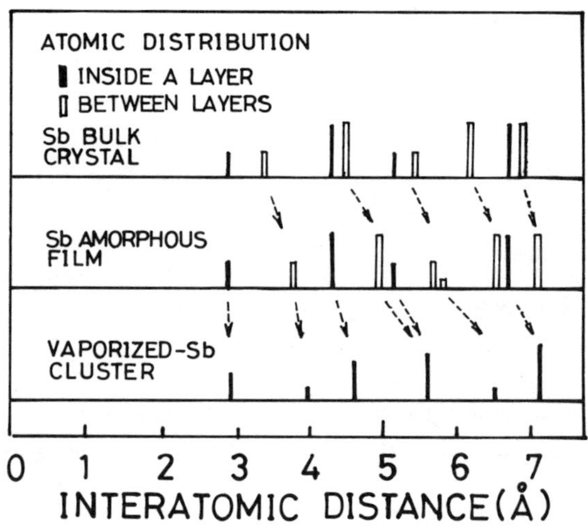

FIG. 12. Comparison of interatomic distances.

The interatomic distance between the nearest-neighbor atoms is larger by 2% than that in crystalline Sb, and the interatomic distances between atoms located farther than the second neighbors are elongated by 7–8% compared with that in crystalline Sb. This result shows that the bond strength in the cluster generated by the vaporized-material ejection might be weaker than that in the crystalline state because of its expanding structure. The weak bonding of atoms in the cluster is an advantage in film formation, because enhanced adatom migration is made possible by momentum transfer from the imported kinetic energy, when the clusters bombard and diffuse on the substrate surface.

III. Film-Formation Mechanism

The use of ionized cluster beams for film formation has the following advantages compared to atomic or molecular ion beams: The ionized cluster is singly charged, i.e., only one atom in a cluster is ionized. Therefore, the beam has an extremely small charge-to-mass ratio, which makes it possible to deposit at a high rate on any kind of substrate, including insulators, with the benefits of the influence of charged ions, but without any problems due to an accumulation of charge. Another advantage involves an inherent property of the clusters, which enables the enhanced migration of adatoms on the substrate surface. Thus deposition at low substrate temperature is possible. Consequently, besides the fundamental effects caused by the kinetic energy and the ionic charge listed in Table I, the above-mentioned characteristic effects can be utilized. Moreover, these effects can be controlled by adjusting the acceleration voltage and the content of the ionized clusters in the total flux.

1. Effect of the Kinetic Energy

In actual film deposition, the ionized clusters are accelerated to an energy range from thermal ejection to a few keV (the energy of the constituent atoms in the ionized cluster is between thermal energy and a few eV). The accelerated clusters produce the following effects: sputtering of surface contamination, surface heating at equivalently high temperature, very shallow ion implantation, formation of preferential nucleation sites, adatom migration, etc.

Some effects can be clearly seen by comparing Si epitaxial growth in ultrahigh vacuum (UHV) and technical-grade high-vacuum chambers. Figure 13 shows diffraction patterns of films deposited under different

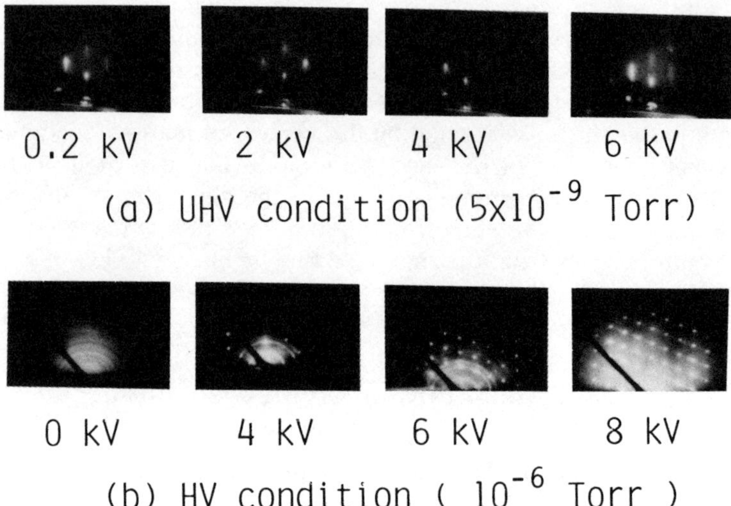

FIG. 13. Electron diffraction patterns of epitaxial silicon films deposited at different acceleration voltages; (a) in an ultrahigh-vacuum (UHV) chamber, and (b) in a technical-grade high-vacuum chamber.

pressure conditions. In the epitaxial growth of Si on an atomically clean and well-ordered silicon surface in the UHV chamber, a 200 V acceleration voltage was enough to obtain epitaxial films on the substrate heated to 500°C. By increasing the acceleration voltage, an improvement in the crystalline quality could be observed. On the other hand, for deposition in a chamber which was evacuated by an oil diffusion pump to a base pressure of 10^{-7}–10^{-6} torr, silicon epitaxial films could be obtained at acceleration voltages higher than 6 kV on a substrate heated to 620°C. An amorphous or polycrystalline structure is formed in the range of 0–4 kV. In this deposition, no special cleaning process, except chemical cleaning, was used prior to the deposition. In this case, a higher acceleration voltage is required in order to sputter the native oxides on the substrate surface and to remove impinging residual gas atoms during the deposition.

Fundamental film formation kinetics by ICB can be analyzed by depositions at different acceleration voltages on substrates heated to various temperatures. The mass deposited, M, expressed by the parameters associated with deposition kinetics, is given by (*23*)

$$M = \dot{M}t - \frac{\dot{M}N_0}{I^*}\exp\left(-\frac{\Phi}{kT}\right) \tag{5}$$

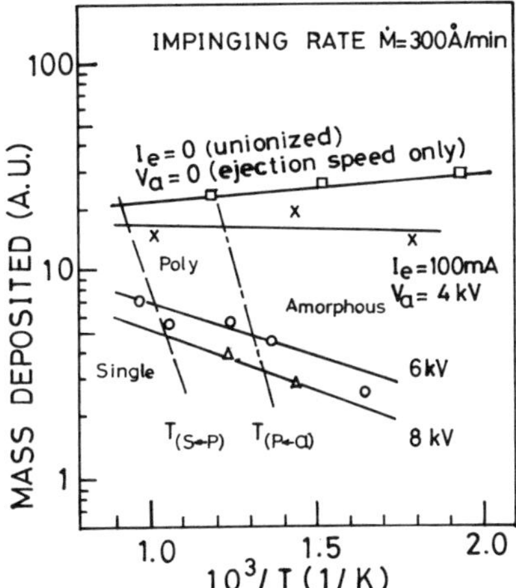

FIG. 14. Characteristics of mass deposited versus reciprocal substrate temperature, and the change of the transition temperature of the crystalline state at different acceleration voltages.

where the energy Φ associated with deposition $= \Phi_{ad} - \Phi_d$, $\dot{M}$ is the mass impinging rate on the substrate, I^* is the rate of formation of the critical nuclei, N_0 is the density of adsorption sites on the substrate surface, Φ_{ad} is the activation energy for desorption, Φ_d is the activation energy for surface diffusion, and t is the deposition time. Figure 14 shows the experimental results for the above relation. The depositions of Si were performed at 10^{-7}–10^{-6} torr on silicon substrates. $T_{(s\leftarrow p)}$ and $T_{(p\leftarrow a)}$, shown in Fig. 14 by the broken lines, indicate the polycrystalline–to–single crystal transition temperatures and the amorphous–to–polycrystalline transition temperatures, respectively, for different acceleration voltages. The transition temperatures $T_{(s\leftarrow p)}$ and $T_{(p\leftarrow a)}$ decrease with increasing acceleration voltage. In the deposition by the un-ionized clusters, the deposited mass increases with decreasing substrate temperature, whereas in the deposition by the ionized clusters the deposited mass decreases with decreasing substrate temperature. Moreover, the slope of the mass-deposited line changes from positive to negative with increasing acceleration voltage. A further increase of the acceleration voltage increases the slope. This may be explained as follows: The decreasing deposited mass with increasing acceleration voltage is due to an increase of the sputtering or re-evaporation of adatoms. The change of slope from positive to negative with increasing acceleration voltage is considered to be due to the change in the values such as N_0, I^*, and Φ. In conventional ion-beam deposition, which uses neutral and ionized atomic particles, the energy Φ does not change (*24*). These experimental results suggest that the fundamental film formation parameters associated with N_0, I^*, and Φ (Φ_{ad} and Φ_d) can be controlled by the change in acceleration voltage. One of the important characteristics of the ICB technique is that the energy associated with deposition is easily controlled by the acceleration voltage.

Figure 15 illustrates the mass impingement rate as a function of substrate temperature and shows the transition temperature $T_{(s\leftarrow p)}$ for silicon films prepared at $V_a = 0$ kV (no acceleration and no ionization) and $V_a = 8$ kV (the case where about 10% of the clusters in the total flux are ionized), respectively. The slope of both lines in the figure yields Φ_d from the following relation (*25*)

$$\dot{M} \leqq A' \exp(-\Phi_d/kT_{(s\leftarrow p)}) \tag{6}$$

where A' is a constant. The result shows that Φ_d decreases with increasing acceleration voltage. This is explained by an enhancement of the surface diffusion energy (migration energy) which could be made possible by increasing the acceleration voltage.

2. Migration Effect

The migration effect was observed by depositing clusters at different acceleration voltages (*26*). The SiO substrate surface was partially covered with a cleaved NaCl plate for shadowing by the edge. The average

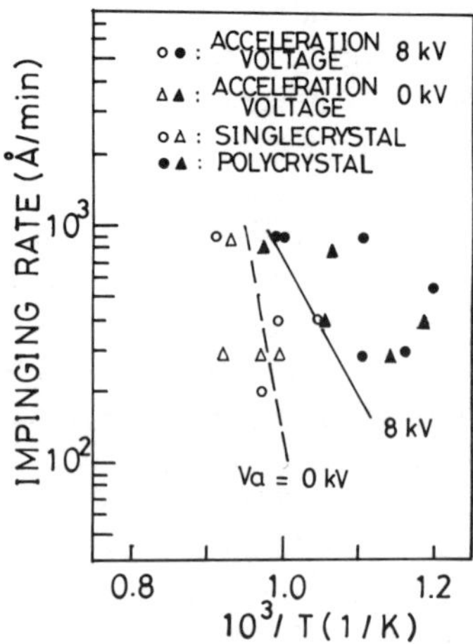

FIG. 15. Characteristics of impinging rate versus reciprocal substrate temperature and the transition temperature from polycrystalline to single crystalline structure.

spacing between the SiO and NaCl mask was about 80 μm. Gold was deposited both by the ICB technique and by conventional vacuum deposition. Figure 16 shows electron micrographs of the deposited Au films near the edge of the mask. In the case of ICB deposition, the deposited Au particles were observed to have migrated under the cleaved NaCl cover. Even when the acceleration voltage was zero, the migration distance of the deposited particles was longer in the ICB deposition than in conventional vacuum deposition. The increased migration distance may be explained by the breaking up of deposited clusters upon impacting the film surface and the imparted energy's being changed to surface diffusion energy.

The average distance $\bar{X}$ transversed over the substrate surface during a time t is given by *(25, 27)*

$$\bar{X} = (4Dt)^{1/2} \tag{7}$$

where D is the diffusion coefficient, and is expressed as

$$D = A \exp(-\Phi_d/kT) \tag{8}$$

where Φ_d is the activation energy for surface diffusion, k is the Boltzmann constant, T is the substrate temperature, and A is a constant. The increase of the migration distance $\bar{X}$ can be related to the decrease of Φ_d by Eqs. (7) and (8), which is caused by the increase in the surface diffusion energy.

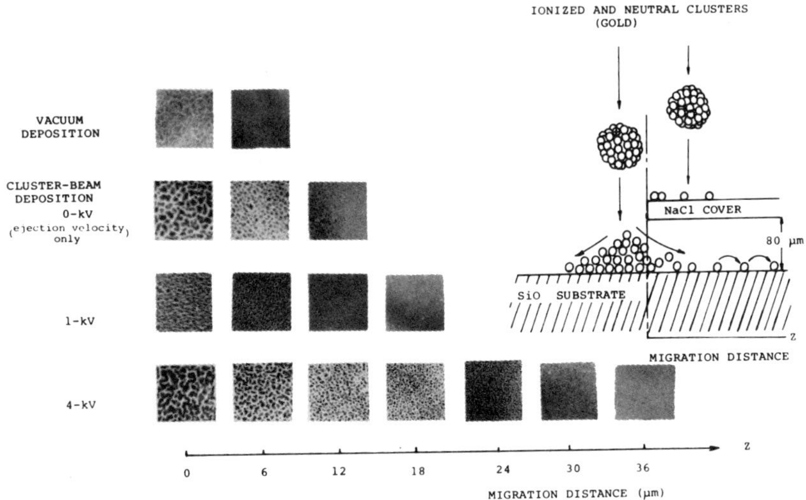

FIG. 16. Electron micrographs of the films formed by migrated deposits.

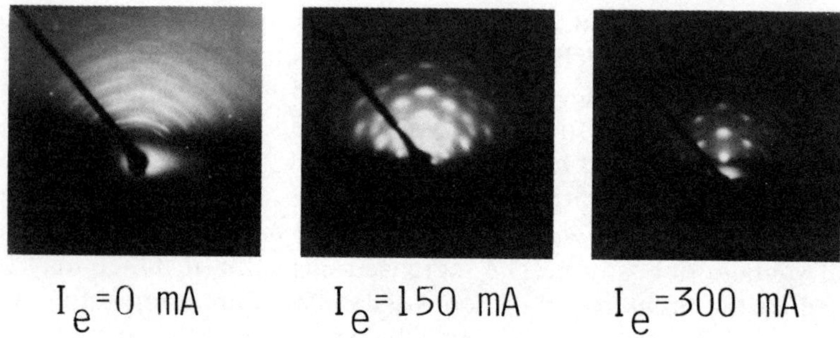

FIG. 17. Influence of electron current (I_e) for ionization on the crystallinity of ZnO films.

3. Effect of the Charged Particle on Film Quality

Effects caused by the presence of the ions, as shown in Table I, can be seen in the change of the critical condensation of the depositing materials and subsequently in the growth mechanism of the nuclei (*23, 24*), even when a few percent of the ionized particles are included in the total beam. The presence of ions is also important to enhance the chemical reaction activity of the evaporated materials. This is useful for forming uniform alloy and compound films (*28*).

In the case of ICB deposition, a typical example of the ionic charge effect was seen in the formation of preferentially oriented ZnO films. The deposition was made without applying acceleration voltages (*29*). Figure 17 shows RHEED patterns of ZnO films deposited onto a glass substrate at different electron currents (I_e) for ionization. The crystallinity of the film is improved as the current is increased. Even in the case of $I_e = 300$ mA, i.e., a degree of ionization of 30–35%, the ion content included in the total flux is 0.03–0.035%, assuming that a cluster contains 1000 atoms. This result demonstrates that the effect of the ionic charge is remarkable, even when only a small amount of ions are included in the beams.

IV. Film Deposition and Epitaxial Growth

Many experiments have been undertaken to prepare various kinds of films by ICB and RICB methods. The films deposited by both methods and their features are summarized in Table III (*30–46*). As shown in the table, the ICB and RICB depositions are available to grow various kinds of films by choosing the appropriate deposition conditions. Some of the characteristics of the films obtained are described here.

TABLE III

FILM PROPERTIES AND THEIR FEATURES

Type of film (application)	Film/ substrate	Acc. Volt. (kV)	Subst. temp. (°C)	Advantages (devices)	Ref.
Metal (coatings, semiconductor metallization)	Au,Cu/glass, kapton®	1–10	RT	Strong adhesion, high packing density, good electrical conduction in a very thin film (high-resolution flexible circuit, optical coating)	(*30*)
	Pb/glass	5	RT	Controllable crystal structure, strong adhesion, smooth surface, improved stability on thermal cycling (superconducting devices)	(*9*)
	Ag/Si(*n*-type)	5	RT	Ohmic contact without alloying	(*31*)
	Ag/Si(*p*-type)		400	Ohmic contact at low temperature (semiconductor metallization)	(*31*)
	AuSb/GaP	2	400	Ohmic contact, strong adhesion (semiconductor metallization)	(*32*)
	Al/SiO_2, Si	0[a]–5	RT–200	Controllable crystal structure, strong adhesion, ohmic contact at low temperature, electromigration resistant (semiconductor metallization)	(*33*)
Intermetallic compound (magnetic films)	MnBi/glass	0[a]	300[b]	*c*-axis preferentially oriented film, spatially uniform magnetic domain, high-density optical memory	(*34*)

TABLE III (*Continued*)

Type of film (application)	Film/ substrate	Acc. Volt. (kV)	Subst. temp. (°C)	Advantages (devices)	Ref.
				(magneto-optic memory)	
	GdFe/glass	0[a]	200	Thermally stable and uniform amorphous (magneto-optic memory)	(*35*)
(Thermoelectric films)	PbTe/glass	0[a]–3	200	Preferentially oriented film, high Seebeck coefficient, low thermal conductivity (high-efficiency thermoelectric converter)	(*36*)
	ZnSb/glass	1	140	Preferentially oriented film, thermally stable, amorphous, high Seebeck coefficient (high-efficiency thermoelectric converter)	(*37*)
	ζ-$FeSi_2$/glass	0[a]–5	150	Thermally stable amorphous, *p*- or *n*-type controllable, high Seebeck coefficient (high-efficiency thermoelectric converter)	(*38*)
Semiconductor (semiconductor devices, EL devices, optoelectronic films)	Si/Si(111) Si/Si(100) Si/sapphire($1\bar{1}02$)	6	620	Low-temperature epitaxy in a high vacuum (10^{-7}–10^{-5} torr), shallow and sharp *p*–*n* junction (semiconductor devices)	(*7*)
	Amorphous Si/ glass	2	200	Thermally stable (solar cell, thin-film transistor)	(*39*)

TABLE III (*Continued*)

Type of film (application)	Film/ substrate	Acc. Volt. (kV)	Subst. temp. (°C)	Advantages (devices)	Ref.
	CdTe/glass	2	250	Improved monocrystalline domains (infrared detector)	(*20*)
	GaAs/GaAs : Cr	6	550	Epitaxy in a high vacuum (10^{-7}–10^{-5} torr) (semiconductor devices)	(*7*)
	GaP/GaP GaP/Si	4	550 450	Low-temperature epitaxy in a high vacuum (low-cost LED)	(*40*)
	ZnS : Mn/NaCl	1	200	Single crystal formation (optical coating)	(*41*)
	ZnS : Mn/glass	1	200	(ac–dc electroluminescent cell with low impedance)	(*41*)
	InSb/sapphire	3	250	Controllable crystal structure (high-mobility devices, magnetic sensor)	(*42*)
Oxide, nitride, carbide, films (coating, functional thin films)	FeO_x/Si,glass	3	250	Controllable optical bandgap (photovoltaic cell, photoelectrode)	(*43*)
	ZnO : Li/sapphire($1\bar{1}02$)	0.5–1	230	*c*-axis preferentially oriented film, single crystal formation (optical waveguide)	(*29*)
	ZnO/glass	0[a]	150	*c*-axis preferentially oriented film, controllable crystal structure and optical transmission (optical waveguide, SAW)	(*29*)
	BeO/sapphire(0001)	0[a]	400	Single crystal formation, transparent film (semiconductor devices,	(*28*)

(*Continues*)

TABLE III (*Continued*)

Type of film (application)	Film/ substrate	Acc. Volt. (kV)	Subst. temp. (°C)	Advantages (devices)	Ref.
				heat sink, electrical insulator)	
	BeO/glass	0[a]	400	*c*-Axis preferentially oriented film, high electrical resistivity and high-thermal-conductivity coating, low-temperature growth (semiconductor devices, heat sink, electrical insulating, and thermal conduction)	(*28*)
	PbO/glass	3	RT	Low-temperature growth, smooth surface, accurately controllable thickness, good adhesion (superconductor)	(*19*)
	GaN/ZnO/glass	0[a]	450	Low-temperature growth of GaN on amorphous substrate (low-cost LED, photocathodic electrode)	(*44*)
	SiC/Si,glass	0[a]–8	600	Controllable crystal state (energy converter, surface protective coating)	(*45*)
Organic material (functional coating)	Anthracene/ glass	0[a]–2	−10–0	Controllable preferential orientation, smooth surface, good adhesion (scintillator, photodiode)	(*46*)

[a] Ejection velocity only.
[b] Annealing temperature.

1. Metal Films

Metal films deposited on an insulating or a metal substrate show good adhesion (*30, 47*). The adhesion strength of a copper film deposited on a glass substrate was measured by a vertical pulling method. The adhesion strength is defined as the force required to peel the deposited film from the substrate. Figure 18 shows the relationship between adhesion strength and acceleration voltage. The adhesion strength increases by a factor of 100 as the applied voltage increases from 0 to 10 kV, which is due to surface cleaning and increased implantation effects.

Metal films with a high packing density can be formed. The packing density is defined by the number of atoms per unit volume in a film. This is generally determined by measurements of film thickness using a stylus, and of the mass per unit area, measured by the electrolytic method. Figure 19 shows the relative packing density for Au films deposited on a copper substrate. The packing density increases with increasing acceleration voltage. The density approached that of the bulk material, when the films were deposited at higher acceleration voltages.

A very thin metal film can be formed. Figure 20 shows the relationship between the resistivity and the thickness of gold films deposited on electron microscope glass plates at different acceleration voltages. As shown in the figure, film resistivity decreases with increasing acceleration volt-

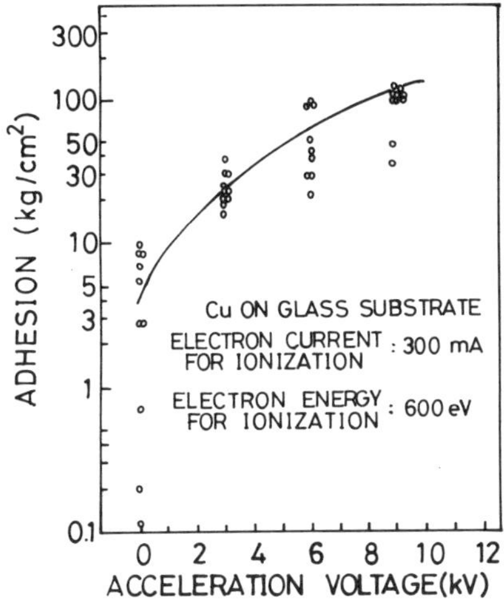

FIG. 18. Adhesion strength of Cu films on a glass substrate as a function of acceleration voltage.

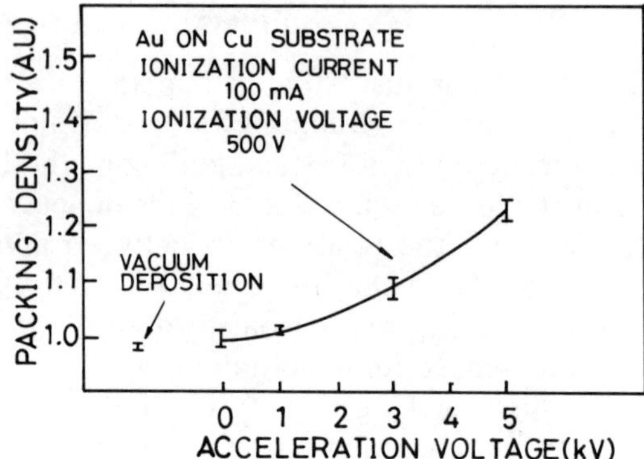

FIG. 19. Dependence of packing density on acceleration voltage for Au films deposited on a Cu substrate.

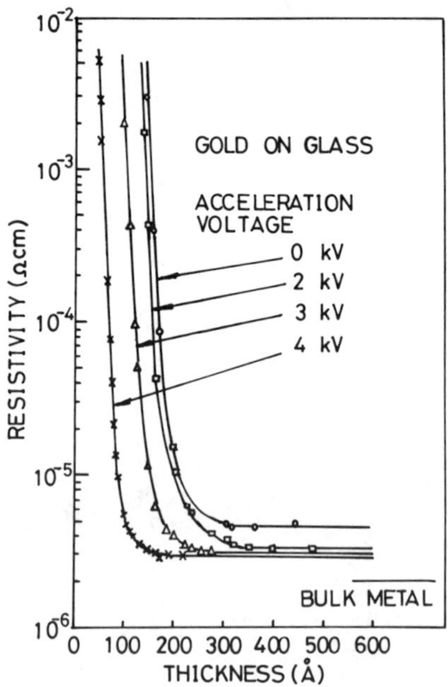

FIG. 20. Resistivity of Au films on a glass substrate as a function of thickness for different acceleration voltages.

age. For the film deposited at an acceleration voltage of 4 kV, the resistivity remains low at a film thickness of 100 Å, even when commercially available glass with ordinary surface roughness is used as a substrate. Although the minimum film thickness to obtain a continuous thin film is determined by a combination of the deposits and the substrate material and by the condition of the depositing surface, the data show the effect of the acceleration voltage clearly. If the film is deposited on an atomically clean surface, an ultrathin film (a few monolayers thick) could be obtained.

Ohmic contacts can be formed at low annealing temperatures. By choosing proper materials, it was found that a multilayer deposition of various kinds of metals, including noble metals, is not always necessary as in the case of vacuum deposition, because ICB permits the introduction of a heavily doped region and/or of numerous recombination centers in the first few monolayers on the semiconductor surface. In Fig. 21, the characteristics of Ag ohmic contacts on *n*-Si substrates are shown. Low contact resistivity was obtained at an acceleration voltage higher than 4 kV without an annealing process (*31, 47*). For a highly doped $p+$ substrate, annealing was not necessary to form an ohmic contact and to get strong adhesion.

Ag deposition by ICB was applied to form the contact electrodes of a *p–n* junction-type solar cell. For comparison, the contact electrodes were made by conventional vacuum deposition in a production line by using the

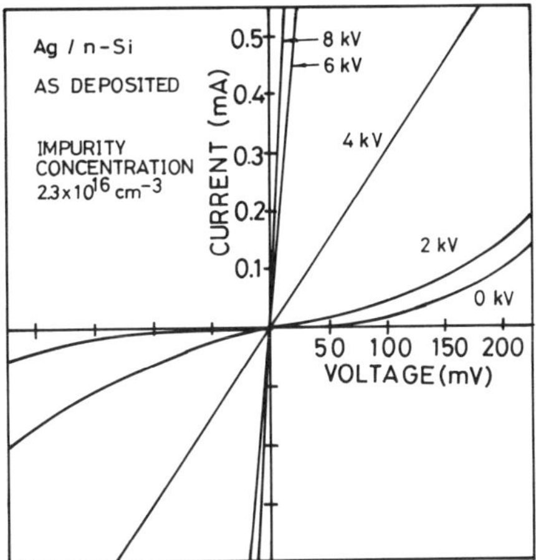

FIG. 21. *I–V* characteristics of Ag contacts on an *n*-Si substrate for different acceleration voltages.

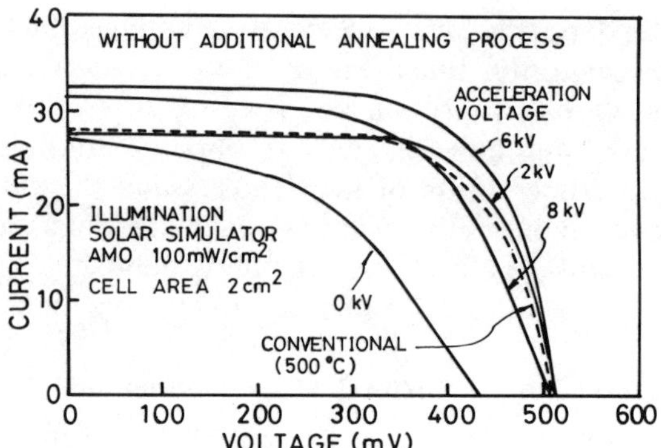

FIG. 22. *I–V* characteristics of a solar cell with the contact electrode deposited by ICB——— and conventional--- evaporation.

same quality *p–n* junction substrate as that for ICB deposition. In the production line the electrode was made by sequential depositions of AuSb, Al, and Ag, followed by sintering at 500°C (*31*). The comparison is shown in Fig. 22. In the case of ICB deposition, deposition at a high acceleration voltage yielded the best electrode characteristics for solar cells without using any sintering process.

Al deposition can be used for VLSI metallization (*33*). The requirements include low-temperature ohmic contact formation, improvement of electromigration lifetime, formation of a smooth surface, and good step coverage. As shown in Fig. 23, an ohmic contact to *p*-Si (10 Ω cm) is

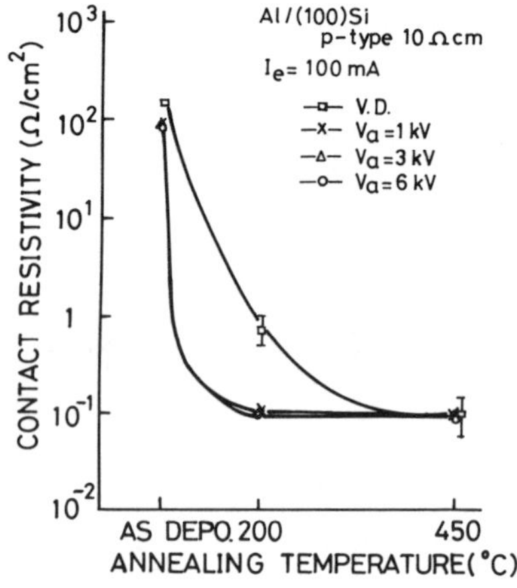

FIG. 23. Contact resistivity as a function of annealing temperature for Al films deposited at different accelerating voltages.

obtained by depositing Al followed by 200°C annealing. In the case of the conventional vacuum-deposited film, 200°C annealing is not enough to obtain low-resistivity contacts. Higher annealing temperatures are needed. For very thin films, electromigration is important. The mean time to failure (MTF) of the interconnection is related to the crystalline size, G, the size distribution of the grain, δ, and the crystalline structure $(I_{(111)}/I_{(200)})$ by

$$\mathrm{MTF} = G/\delta^2 \ln(I_{(111)}/I_{(200)})^3 \tag{9}$$

where $I_{(111)}$ and $I_{(200)}$ are the x-ray diffraction intensity values for the (111) and (200) planes. From x-ray diffraction measurements, an increase of acceleration voltage results in an increase of the ratio of $I_{(111)}$ to $I_{(200)}$, as shown in Fig. 24. The size of the grains becomes larger at higher acceleration voltages. A narrow distribution of the grain size was also observed. Therefore, these factors contribute to increase the lifetime when the deposition is made at higher acceleration voltages. Uniform coverage over variously shaped steps is obtained, as shown in Fig. 25

2. Magnetic Films

a. MnBi Films. Uniaxial MnBi films with an easy axis perpendicular to the surface of a glass substrate possess many unique magnetic properties, and are suitable for high-bit-memory applications for a Curie-point writing

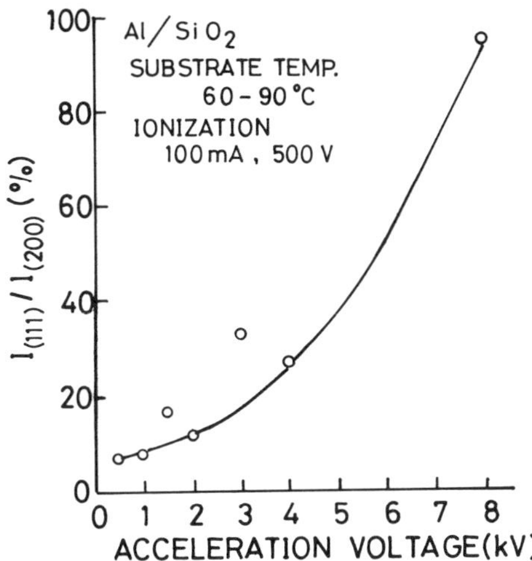

FIG. 24. Ratio of x-ray diffraction intensity of the (111) plane to that of the (200) plane for Al films deposited at different acceleration voltages.

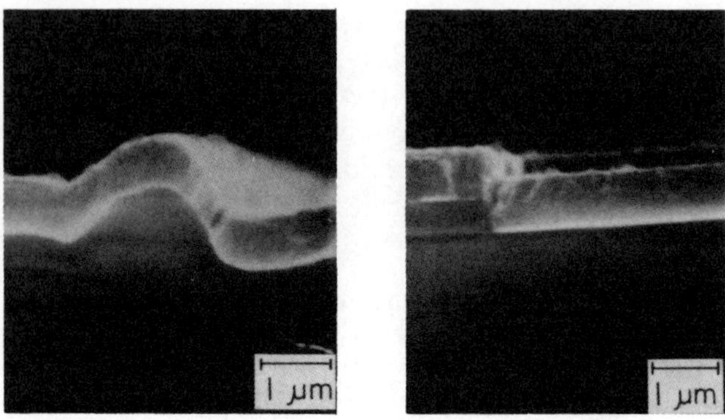

FIG. 25. SEM images of Al films deposited on differently shaped steps.

and magneto-optical readout devices using a laser. Entirely uniform films over large substrate areas, which were difficult to prepare by conventional methods, have been obtained easily by a two-crucible version of the ICB technique, where Mn and Bi were deposited simultaneously on glass substrates (*34*). The crystalline easy axis of the films was oriented in the direction perpendicular to the substrate, and could be aligned preferentially as ion contents were increased.

Domain patterns were observed with the aid of a polarizing microscope. Figure 26a and b shows the typical magnetic domain structure of MnBi films deposited on glass substrates, followed by annealing at 300°C for 3 h, in comparison with that of a MnBi film deposited by a conven-

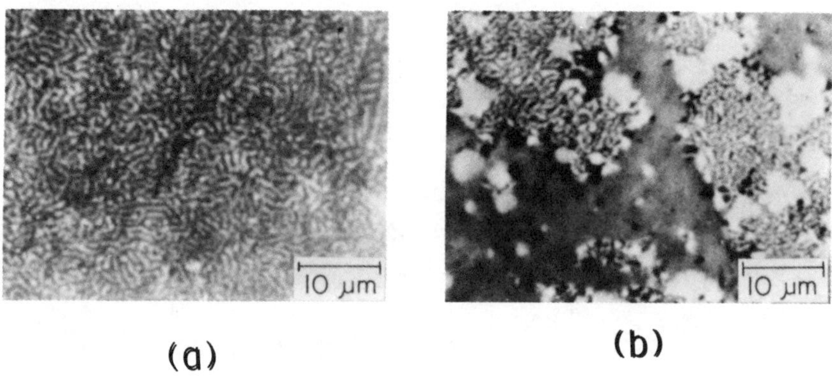

FIG. 26. Magnetic domain configurations of MnBi films prepared by (a) the ICB deposition technique and (b) conventional vacuum deposition.

tional co-evaporation technique without ions. As seen in Fig. 26a, uniaxial stripe domains with the easy axis perpendicular to the film surface appear all over the glass substrate (5×5 cm^2). On the contrary, for conventional vacuum evaporation, as shown in Fig. 26b, islandlike domain structures grow, and the remaining part consists of nonmagnetic materials such as MnO and Bi_2O_3.

Measurements of the Faraday rotation at the He–Ne laser wavelength (6238 Å) were made for films deposited at different values of I_e, which were magnetically saturated by applying 6 kOe in a direction perpendicular to the film surface. Figure 27 shows the Faraday rotation versus annealing temperature. Chen's results (*48*) for double layers of Mn and Bi evaporated on a mica substrate are compared with the present results. For the films deposited using the ICB technique, the Faraday rotation rises at annealing temperatures below 80°C, which is about $\frac{1}{3}$ of that for the conventional evaporation. A pronounced increase in the rotation angle can be seen for higher amounts of ionized clusters.

The material's characteristics as a magneto-optical memory were evaluated by Curie-point writing using a He–Ne laser. The high stability of the film was maintained after the writing and erasing process over a substantial number of cycles, which proves that the adhesion strength of the film to the substrate is high.

b. GdFe Films. Amorphous magnetic films of a GdFe binary alloy have been investigated to develop a thermally stable memory device capable of high-bit addressable memory with a semiconductor laser. Preparation of GdFe thin films was by simultaneous deposition of Gd and Fe from separate crucibles (*35*). From Auger analysis on GdFe films, they were found to have a uniform stoichiometric composition over a large surface area (5×5 cm^2).

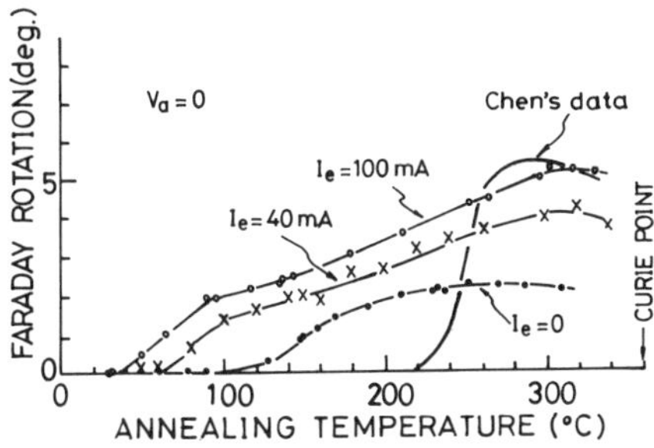

FIG. 27. The MnBi growing process during annealing as measured by the magneto-optical Faraday effect.

From measurements of the x-ray radial distribution function (RDF), Gd–Fe pair bondings around 3.2 Å in $G(r)$ were dominant even when 30% of the Fe clusters were ionized, and the Fe–Fe pair density [about 2.5 Å in $G(r)$] was somewhat smaller than that of films made by sputter deposition.

From FMR experiments, the internal effective magnetic field H_{eff} of as-grown GdFe films was found to be negative, indicating its magnetization direction was in the plane of the film. Since $H_{\text{eff}} = 2K_u/M_s - 4\pi M_s$ (K_u, the uniaxial anisotropy coefficient; M_s, the saturation magnetization), the value of K_u must be large or that of M_s must be reduced to obtain amorphous GdFe films having a perpendicular magnetization.

The internal effective magnetic field H_{eff} of a Bi (20 at. %)-doped GdFe film was larger than that of pure GdFe films, and was positive ($H_{\text{eff}} =$ 147.8 Oe) after annealing at 300°C for 2 h in a vacuum chamber. From the results of RDF with respect to Bi-doped GdFe films, it was found that the perpendicular magnetization of the films was responsible for an increase of Gd–Bi pairs and a decrease of Gd–Fe direct pairs, which is responsible for the reduction of the saturation magnetization M_s.

3. Thermoelectric Films

a. PbTe Films. For preparation of PbTe films using the ICB technique a two-crucible system was used, in which the materials were ejected from separate crucibles (*36*). Only tellurium clusters were ionized by an electron current (I_e) and accelerated toward the substrate by adjusting the acceleration voltage (V_a). Lead clusters were neutral, because the ionized cluster beam of tellurium more effectively resulted in growth along the <200> axis of PbTe than did ionizing the lead clusters. The substrate temperature was kept constant at about 200°C during deposition.

Figure 28 shows the x-ray diffraction patterns of PbTe films. The film was deposited at $V_a = 3$ kV by changing I_e. The enhancement of the preferential orientation along the <200> axis is seen with increasing I_e. PbTe films were also deposited at different values of the acceleration voltage, keeping I_e constant at 300 mA, and higher acceleration voltages were found to enhance the film growth along the <200> axis.

Measurements of the thermoelectric power (the Seebeck coefficient) S were made for PbTe films deposited at $I_e = 300$ mA and $V_a = 0$–3 kV. The values of S obtained were about 500 μV/deg for *p*-PbTe films and -300 μV/deg for *n*-PbTe films, which are found to be 3–4 times larger than those of PbTe ingots (bulk) prepared by the conventional thermal method or the Bridgman technique (*49*). This is ascribed to uniaxially

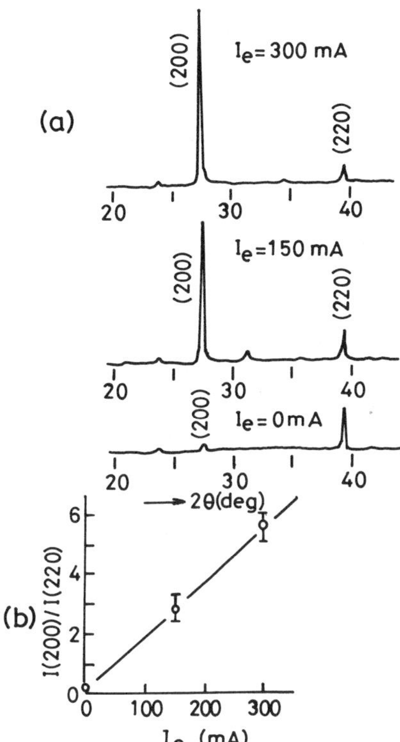

FIG. 28. (a) X-ray diffraction patterns and (b) the ratio of intensity from the (200) plane to that from the (220) plane for PbTe films deposited on glass substrates at different electron ionization currents.

oriented film growth aided by the presence of ions and their kinetic energies.

b. ZnSb Films. The compound ZnSb has been investigated as a *p*-type thermoelement, which is used in the range from room temperature to 450°C. ZnSb has an orthorhombic structure, which is regarded as a strongly deformed diamond structure (*50*). Zn–Sb composite films were prepared by the ICB technique, and the crystallographic and thermoelectric properties were measured (*37*). It was possible to prepare crystalline or amorphous films without a peritectic reaction at a lower temperature than the melting point of ZnSb in the presence of ions and with the kinetic energy of the clusters.

The source materials were put in each crucible separately and were heated to obtain a stoichiometric composition. ZnSb films were deposited on a glass substrate by changing the electron current (I_e) for ionization and the acceleration voltage (V_a). The pressure during deposition was 5×10^{-6} torr, and the substrate temperature (T_s) was in the range of 100–150°C.

Thermoelectric properties of crystalline films with preferential orientation along the <102> axis were measured in the case when Zn clusters were ionized and Sb clusters were neutral. The deposition conditions were: $I_e = 300$ mA and $V_a = 1$ kV. The temperature dependence of the Seebeck coefficient (S) showed a maximum value of 600 μV/deg at 550 K at the stoichiometric composition. The figure of merit Z, defined by $Z = S\sigma^2/\kappa$, for the film obtained was compared with that of a polycrystalline specimen prepared by conventional thermal melting. (Here, σ and κ are the electrical and thermal conductivity, respectively.) The value of Z for the ZnSb crystalline film was of the order of 1×10^{-2} K^{-1} at 520 K, larger than the maximum value obtained for a polycrystalline bulk specimen (1.25×10^{-3} K^{-1} at 400 K).

By the ICB technique, ZnSb amorphous films could be obtained easily by adjusting acceleration voltages and substrate temperatures while keeping I_e constant. The temperature dependence of the electrical conductivity for amorphous ZnSb films deposited under different conditions is shown in Fig. 29. The deposition conditions were: $I_e = 300$ mA, $V_a = 1$ kV, and $T_s = 80$°C. It was found that there are two transition temperatures (T_{t1} and T_{t2}) at which one can recognize the change from the amorphous to the crystalline state and a major increase in σ. From the slopes in the temperature range in which an amorphous structure exists, the activation energies are estimated to be 0.36 eV for sample 1, 0.23 eV for sample 2, and 0.28 eV for sample 3. These results are ascribed to the effect of ions on the film formation.

c. $FeSi_2$ Films. Amorphous films of an iron disilicide (ζ-$FeSi_2$) were prepared by the ICB deposition technique (two-crucible system) to develop thin-film energy-conversion devices (*38*).

For *p*-type films, Fe and SiO were used as source materials and were heated in separate carbon crucibles. Only Fe clusters were ionized. The electron current (I_e) for ionization was 200 mA. The substrate temperature was kept at about 150°C. From the results of analysis of these films by XPS (x-ray photoemission spectroscopy), *p*-type thermoelectric power could be ascribed to Si–O pairs in the $FeSi_2$ films during deposition.

Figure 30 shows the temperature dependence of the thermoelectric power (the Seebeck coefficient, S) of *p*-type Fe–Si films with a composition in the range 50–86 at. % Si, which have amorphous structures. The Seebeck coefficient of ε-FeSi, consisting of 50 at. % Si (sample 50) is of the order of +40 μV/deg. For ζ-$FeSi_2$, consisting of 72 at. % Si (sample

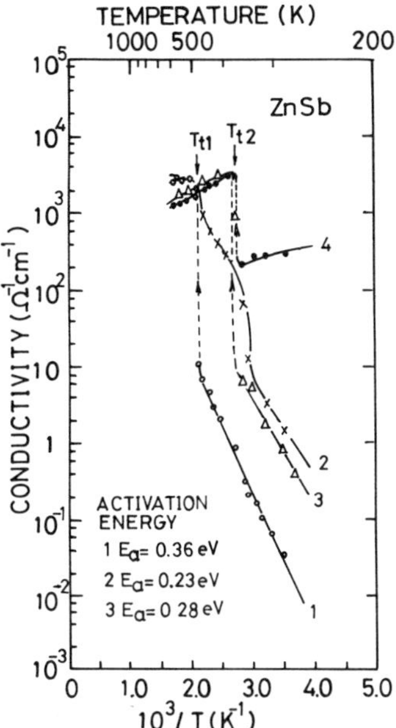

FIG. 29. Electrical conductivity of ZnSb amorphous films: 1, only Zn clusters ionized; 2, only Sb clusters ionized; 3, both Zn and Sb clusters ionized; 4, both Zn and Sb clusters neutral.

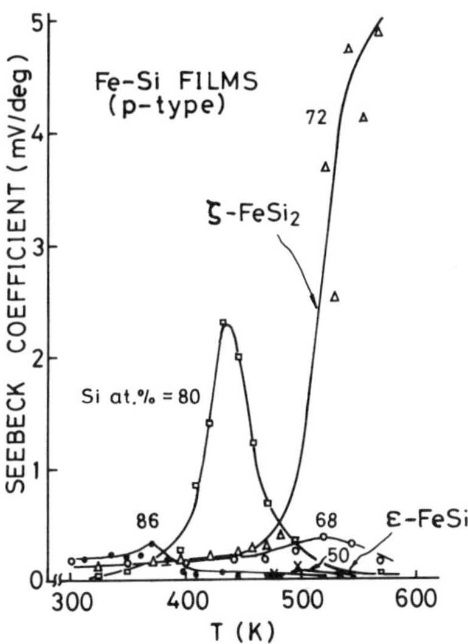

FIG. 30. Temperature dependence of the Seebeck coefficient for *p*-type iron silicide films.

72), an extremely high value of +5000 μV/deg is obtained at about 550 K. With a Si content beyond the region of the ζ-$FeSi_2$ phase, the Seebeck coefficient decreases. For example, $S = +2200$ μV/deg for 80 at. % Si (sample 80), and $S = +300$ μV/deg for 86 at. % Si (sample 86). Thus, the peak value of S obtained for the film is found to shift toward the lower-temperature region with increasing Si content. Further, as shown in Fig. 31, the peak value of S for the $FeSi_2$ films changes with the number of Si–O pairs incorporated in the film. The value is found to be about 5 mV/deg at 550 K in the case of 13% of Si–O pairs (i.e., $FeSi_2:0_{0.26}$), and to be about 10 mV/deg at 650 K when almost all the Si pairs with O atoms (i.e., $FeSi_2:O_{1.87}$).

Type conversion was made possible by the RICB technique using the reaction between vaporized Fe cluster ions and oxygen. n-type thermoelectric power can be obtained by Fe–O pairs in the films. Figure 32 shows the temperature dependence of the Seebeck coefficient S for an amorphous, n-type, ζ-$FeSi_2$ film containing 70 at. % Si, deposited at an oxygen partial pressure of 8×10^{-5} torr. The conduction type is re-

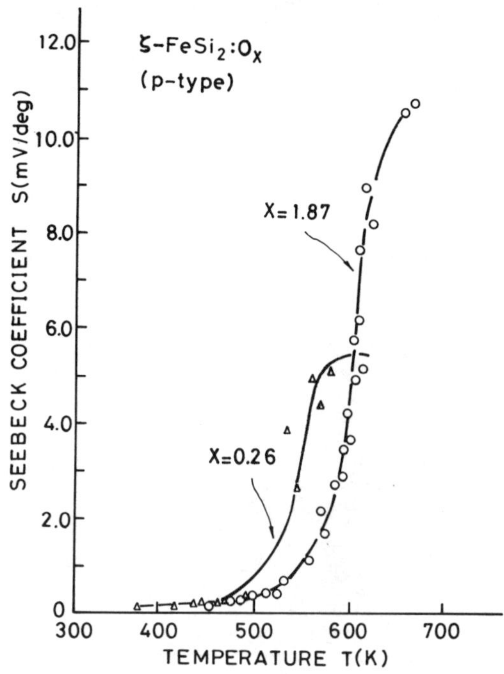

FIG. 31. Temperature dependence of the Seebeck coefficient for p-type iron disilicide films. Oxygen content: (a) $X = 0.26$ (△) and (b) $X = 1.87$ (○).

versed at a transition temperature near 400 K, and the value of S increases markedly with increasing temperatures. An extremely high value ($S = -20000\ \mu V/deg$) is obtained at about 580 K. The transition temperature can be reduced with increasing acceleration voltage.

4. Semiconductor Films

a. Silicon Epitaxial Growth. Si epitaxial films can be formed by ionized cluster-beam deposition at substrate temperatures lower than 620°C at a pressure of 1×10^{-7} torr. A cluster source was heated to 2000–2200°C. High-density carbon of semiconductor grade was used as a crucible material. The ionization current was 100 mA. Prior to the deposition, the Si substrate, with a (111) orientation surface, was chemically cleaned with trichloroethylene, hydrofluoric acid, and acetone. Epitaxial growth could be achieved at an acceleration voltage of 4–8 kV on substrates heated in the temperature range 300–620°C, as previously shown in the lower part of Fig. 13. This epitaxial temperature is well below that required during vacuum evaporation in a UHV chamber (*16*). Figure 33 shows the comparison of ion microanalysis of (a) the Si epitaxial film deposited by using the carbon crucible and (b) a CZ-grown Si wafer which is commercially available. The purity of the deposited film is found to be of the same order as that of the CZ-grown Si wafer. Figure 34 shows typical backscattering

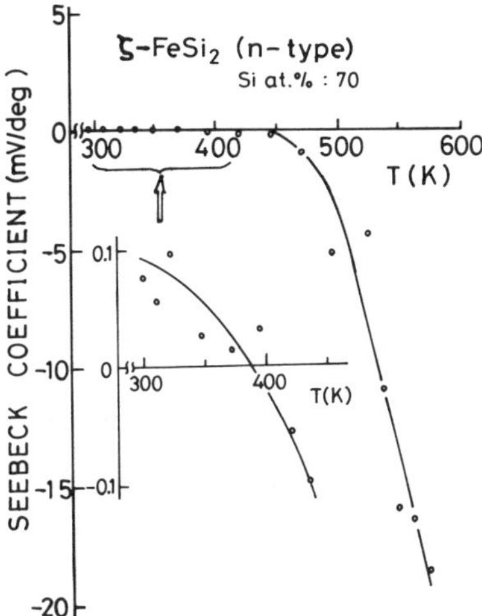

Fig. 32. Temperature dependence of the Seebeck coefficient for n-type iron disilicide films.

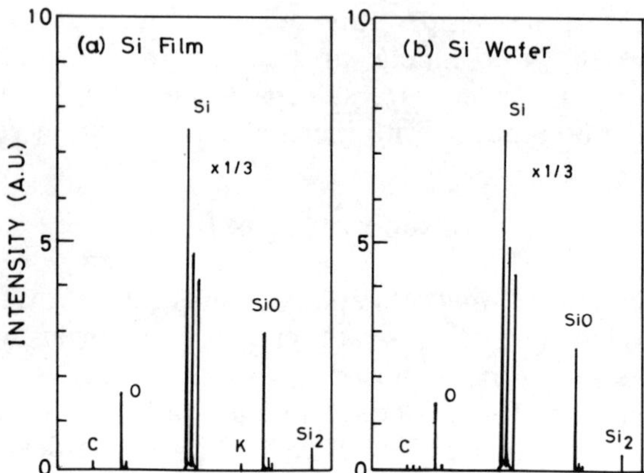

FIG. 33. Comparative ion microanalysis of (a) Si film deposited on a Si substrate by ICB and (b) CZ-grown Si wafer.

energy spectra of the Si film deposited at 620°C, measured by 185 keV H^+ ions. In the spectrum, no peak corresponding to the oxygen could be observed. The minimum yield, χ_{min}, which is expressed as the ratio of the aligned yield to the random yield just below the surface peak, was about 6.7% (*51*). The χ_{min} of the substrate used in this experiment was between 4 and 7%.

Electrical properties of the *n*-type films (1300–1500 Å thick) deposited on *p*-type substrates were measured by the van der Pauw method. The

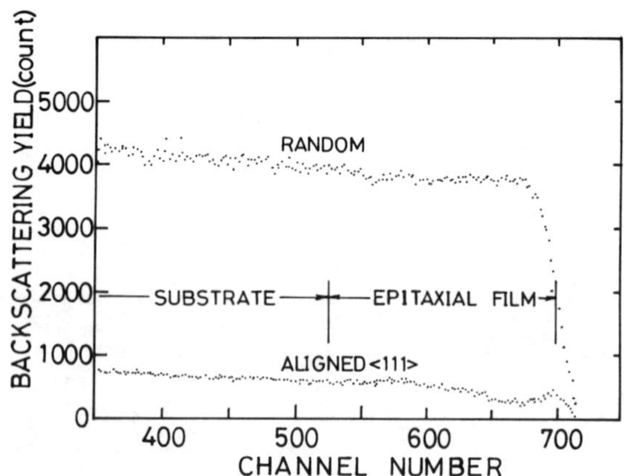

FIG. 34. Rutherford backscattering analysis of a Si epitaxial film grown by ICB deposition at an acceleration voltage of 6 kV on a Si(111) single crystal at 620°C.

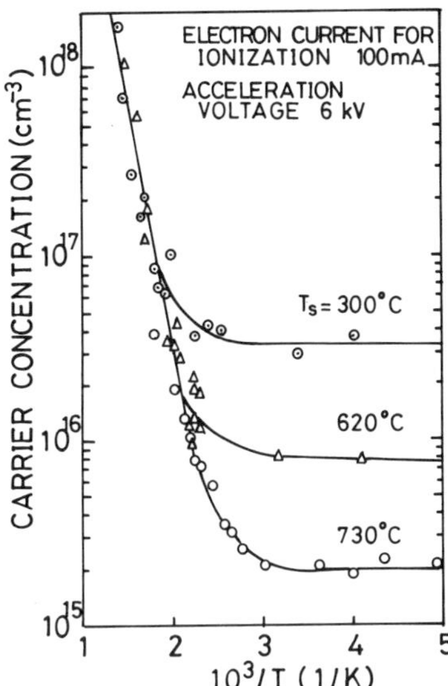

FIG. 35. Temperature dependence of negative-carrier concentrations for Si films grown at different substrate temperatures. T_s = 300 (⊙), 620 (△), and 730°C (○).

Hall mobility increased with an increase in accelerating voltage and with an increase in substrate temperature. The mobility of the film grown at an acceleration voltage of 8 kV and a substrate temperature of 620°C was in the range of 700–800 cm^2/V sec (*31*).

Figure 35 shows the temperature dependence of the carrier concentration in the films deposited at three substrate temperatures, 300, 620, and 730°C. The intrinsic region is seen in all the samples measured at high temperatures. At lower temperatures, the donor-exhaustion region is clearly seen; the carrier concentration does not change appreciably with temperature. This suggests that good-quality Si films can be obtained. The carrier concentration is higher for films deposited at low substrate temperatures than for films deposited at higher temperatures. This is probably due to the difference in the doping concentration, which is caused by the different sticking coefficients of the donor impurities during deposition.

p–n junction diodes were fabricated by the deposition of *n*-type Si films on *p*-type single crystal Si substrates at different deposition temperatures. The *C–V* curves of the diodes are shown in Fig. 36. The curves show that the samples deposited at 620 and 300°C have an abrupt junction, while for the film deposited at 730°C the junction characteristics were slightly degraded. The degradation might be due to the introduction of impurities

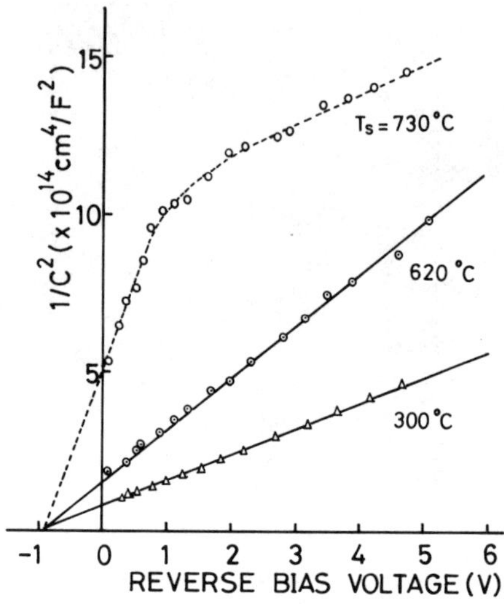

FIG. 36. *C–V* characteristics of a *p–n* junction Si diode measured at 1 kHz. T_s = 300 (△), 620 (⊙), and 730°C (○).

during the deposition. These results suggest that it is possible to fabricate devices with abrupt junctions because of the low temperature of epitaxial growth. A *p–n* junction solar cell was fabricated, and the photovoltaic characteristics were measured. The junction was made by depositing *n*-type Si films onto a *p*-type Si [6–11 Ω cm, (111) orientation] substrate with varying film thicknesses. The spectral characteristics of such photovoltaic diodes are shown in Fig. 37. The results show that epitaxial films with the desired thickness could be obtained, which may be applicable to nanometer-structure device fabrication.

b. Amorphous Silicon Films. Doped *a*-Si : H films with smooth surfaces, strong adhesion, and thermally stable characteristics can be formed by the ICB technique (*39*). The structure of Si films could be changed from the amorphous state to a single crystal by changing the acceleration voltage and the substrate temperature, as shown in Fig. 38. These experimental data suggested the possibility of forming amorphous Si films with controlled film characteristics. In conventional fabrication methods, such as glow discharge deposition or reactive sputtering, the film is deposited from a plasma in the gas pressure range 10^{-2}–a few torr. Compared with the conventional method, in the ICB method the film is formed in a hydrogen pressure as low as 10^{-4} torr. Figure 39 shows the dependence of the optical bandgap of the films on the hydrogen gas pressure. The films

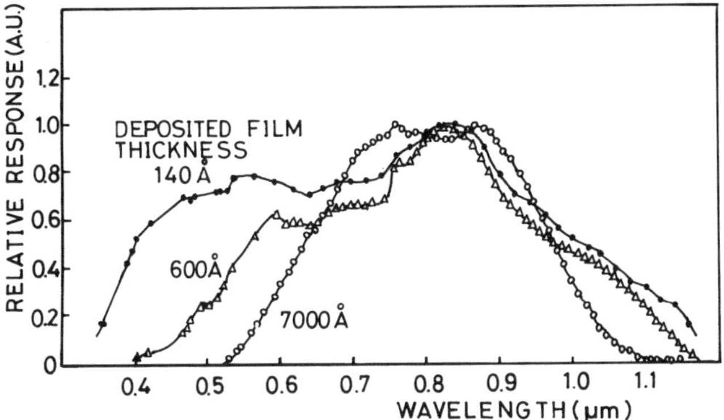

FIG. 37. Open-circuit photovoltage versus wavelength characteristics for different thicknesses of *n*-Si films on *p*-Si substrates.

were deposited at 3 kV on glass substrates heated to 220°C. A change in the optical bandgap of 1.3–1.9 eV could be obtained by changing the hydrogen gas pressure from 0 to 3×10^{-4} torr. Figure 40 shows the change of the optical bandgap of the films deposited at different acceleration voltages. The optical bandgap increases with increasing acceleration voltage. The data show that the hydrogenation of the films can be enhanced at higher acceleration voltages. Annealing characteristics were

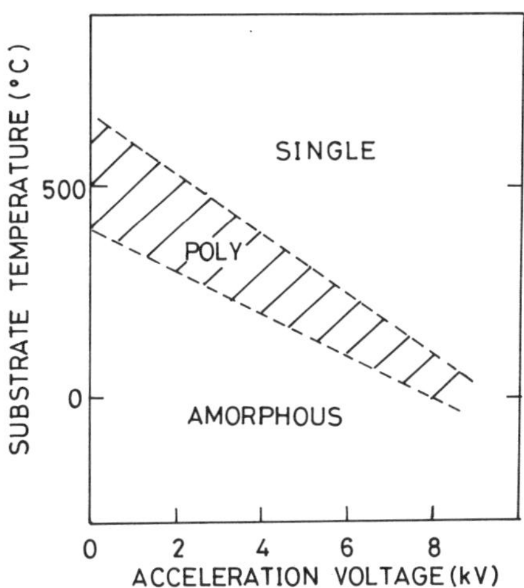

FIG. 38. Acceleration voltage and substrate temperature dependence of the crystallinity of silicon films deposited by ICB.

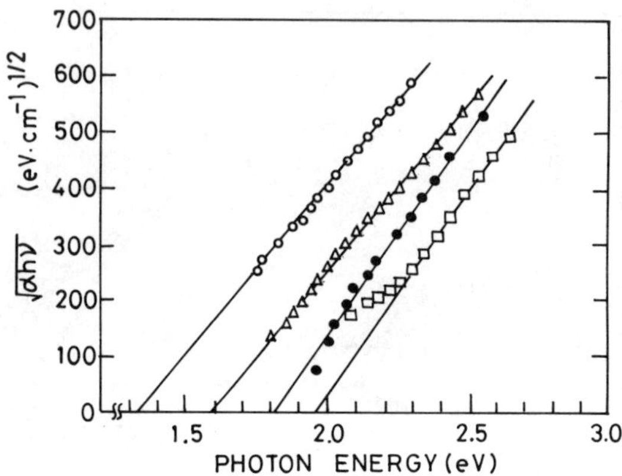

FIG. 39. Dependence of the optical bandgap for *a*-Si : H films on various hydrogen pressures: (○), 0; (△), 3×10^{-5}; (●), 1×10^{-4}; and (□), 3×10^{-4} torr.

measured with films deposited at different acceleration voltages. The optical bandgap of the films deposited at higher acceleration voltages does not change after annealing at 400°C in vacuum for 30 min. This result shows that the formation of thermally stable films is possible. The infrared absorption of the films deposited at different acceleration voltages was measured. As shown in Fig. 41, the inclusion of monohydrides at 2000 cm^{-1} increases with increasing acceleration voltage, whereas the inclusion of

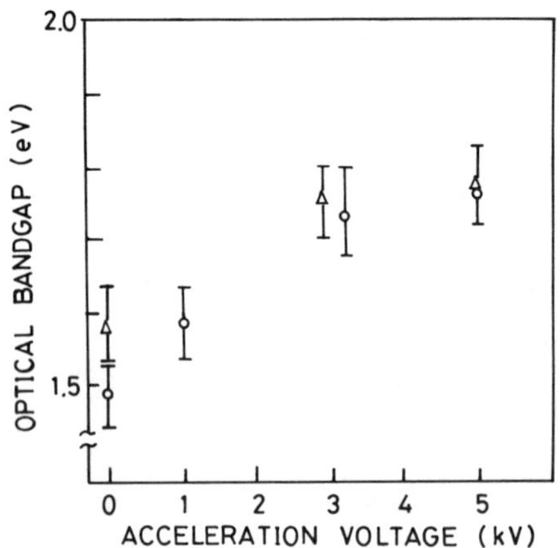

FIG. 40. Optical bandgap of *a*-Si : H film as a function of the acceleration voltage. (○), As-deposited film; (△), annealed at 400°C.

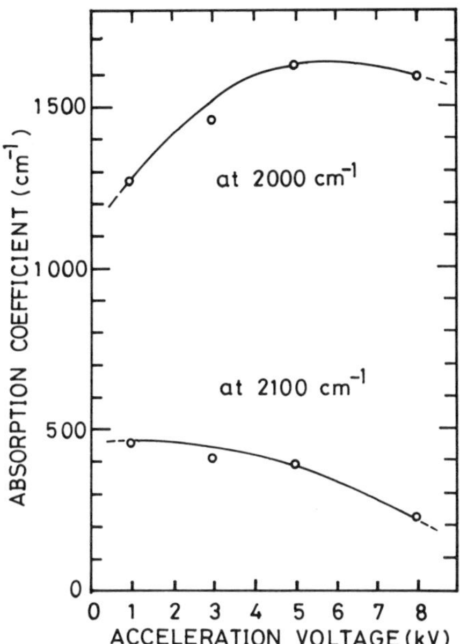

FIG. 41. Absorption coefficients at 2000 and 2100 cm^{-1} for *a*-Si:H films deposited at different acceleration voltages.

dihydrides (corresponding to a 2100 cm^{-1} wave number) decreased. The uniform hydrogenation under a very low hydrogen gas pressure could be explained effective and uniform chemical reaction due to enhanced adatom migration and the presence of ionic charge. Although the thermal stability of the films is mainly due to the structural properties of the dominant monohydride, another reason may be that so-called gas precipitation or inclusion of gases is extremely small, because the physically absorbed hydrogen gases are easily desorbed by the bombardment of accelerated ions. Electrical characteristics such as photo and dark conductivities are also improved by deposition at higher acceleration voltages, as shown in Fig. 42. For doped film formation hydrogen gas mixed with phosphine or diborane (of the order of 5000 vppm) was used. A *p–i–n* diode could be fabricated on a metal substrate sequentially by changing doping gases. No trace of cross-contamination by doping gases in each layer was observed.

c. ZnS : Mn Films. Most of the work on thin-film dc electroluminescent (EL) cell development was concentrated on ZnS doped with Mn, Cu, or Cl (*52*). dc EL in ZnS : Mn by simultaneous or alternating implantation

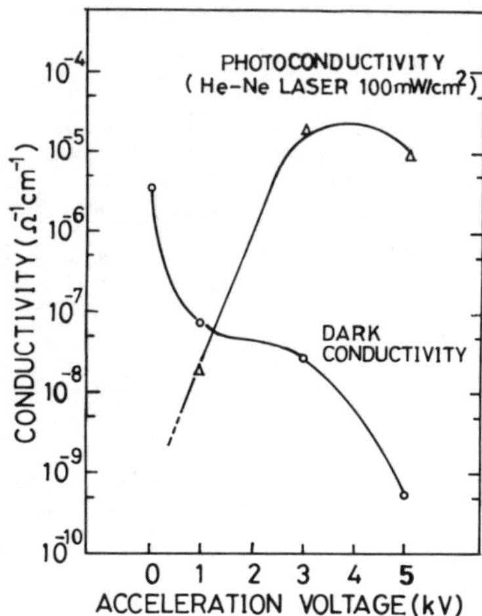

FIG. 42. Photoconductivity and dark conductivity for *a*-Si : H films as a function of the acceleration voltage.

and vacuum deposition was reported by Takagi *et al.* (*53*) in 1973. The cells showed long life under dc operation. By using the ICB deposition, similar results have been obtained.

In the experiments using the ICB technique, ZnS : Mn was deposited at 1 kV with a substrate temperature of 250°C (*41*). After deposition, the film was annealed at 500°C for 1 h in a vacuum, and then Al was deposited as a contact electrode. Figure 43 shows the electron diffraction pattern of the ZnS : Mn film deposited on a rocksalt substrate. The single crystal state is indicated by the pattern shown in this figure.

The current–voltage and the voltage–brightness characteristics of the film deposited on NESA glass by the ICB technique were investigated.

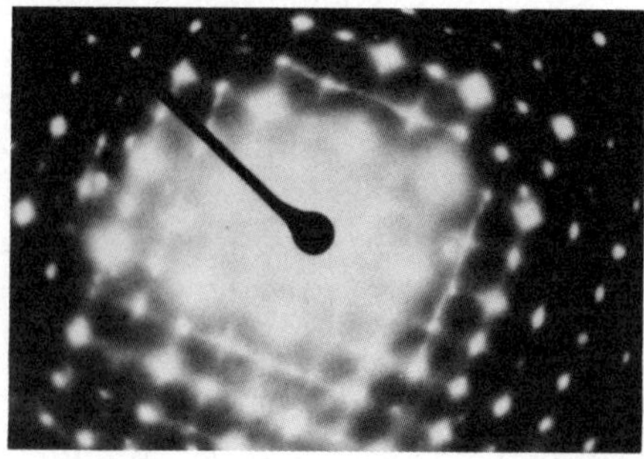

FIG. 43. HEED pattern of ZnS : Mn film prepared by ICB.

The cell showed low impedance characteristics under dc excitation similar to those of a ZnS : Mn cell with a coactivator made by a conventional method, but it did not show the rectification characteristics which are exhibited by a dc EL cell fabricated by thermal diffusion processing.

d. GaP Films. Epitaxial films of GaP on GaP and Si substrates can be grown at low substrate temperatures by the ICB technique (*40*). In the homoepitaxial growth of GaP, an *n*-type GaP substrate was used, and the substrate temperature was 550°C. The films deposited with neutral cluster beams of Ga and P were polycrystalline. When the Ga clusters were ionized and P clusters remained neutral, the crystalline state of the films was improved. Further, when both Ga and P clusters were ionized, the film became single crystal, and the surface obtained was found to be smooth from streaked RHEED patterns.

The lattice constant of GaP is close to that of Si (misfit: 0.37%), but the thermal expansion coefficient of GaP is about twice that of Si. Therefore, in the heteroepitaxial growth of GaP on Si, growth at low substrate temperatures is desired. Epitaxial films grown with ionized Ga and P clusters were obtained at an acceleration voltage of 4 kV and with a substrate temperature as low as 450°C.

p–n junction LEDs were fabricated by depositing *p*-type GaP on *n*-type GaP substrates. The cell emitted light from about 15 to 20 mA (forward bias) at room temperature. The costs of LEDs fabricated by the ICB technique are expected to be much lower than those using other techniques.

5. Oxide, Nitride, and Carbide Films

a. ZnO Films. ZnO films have the wurtzite structure and are a dielectric material having a bandgap of 3.2 eV and a high electromechanical coupling coefficient ($k_{em} = 0.4$). The material attracts interest for optoelectronic and surface-acoustic-wave devices. ZnO films have been prepared by the RICB technique (*29*). The acceleration voltage (V_a), the electron current (I_e) for ionization, and the substrate temperature (T_s) were: $V_a =$ 0–1 kV, $I_e =$ 0–300 mA, and $T_s =$ 100–250°C. Oxygen gas was introduced into the chamber at a partial pressure of 5×10^{-4} torr.

From the results of x-ray diffraction and RHEED patterns of films deposited on glass substrates under various conditions, it was found that the films were preferentially oriented along the *c* axis at $V_a =$ 0–500 V and $I_e =$ 300 mA. The preferential orientation increased as the electron current for ionization increased, which was previously shown in Fig. 17. The smoothness of the films deposited at a higher acceleration voltage

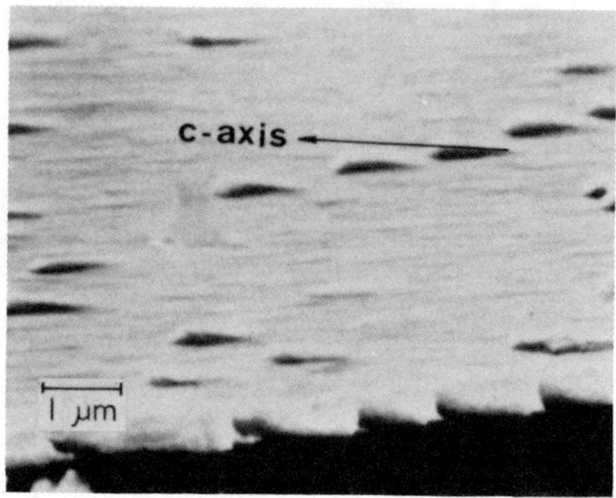

FIG. 44. SEM view of ZnO film grown on a $(1\bar{1}02)$ sapphire substrate.

was better, and the attenuation loss of the films could be changed by the acceleration voltage.

Single crystal films could be obtained on different planes of sapphire substrates. In particular, the films grown on $(1\bar{1}02)$ sapphire substrates at $V_a = 1$ kV and $I_e = 300$ mA are found to be preferentially oriented with their c axis almost parallel to the substrate, as shown in Fig. 44.

b. BeO Films. BeO crystallizes in the wurtzite structure and is an insulator with high resistivity (10^{11} Ω cm at 300 K) and with a large bandgap (11.2 eV) (*54*). In addition, the material has a high thermal conductivity (of the same order as that of Al) resulting from lattice vibrations. Other interesting properties are high mechanical hardness (Mohs' hardness-9) and excellent chemical stability.

c-axis preferentially oriented BeO films on glass substrates and good-quality epitaxial films on sapphire (0001) substrates were grown using RICB under the following deposition conditions: acceleration voltage V_a was 0, electron current (I_e) for ionization was 300 mA, substrate temperature (T_s) was about 400°C, and the oxygen pressure was 3×10^{-4} torr. The films obtained were completely transparent, and the crystallinity and surface smoothness were preeminent.

In Fig. 45, an x-ray diffraction pattern of a BeO film grown on a glass substrate is compared with that of a BeO ceramic. In this figure, the SEM structure of the BeO film, showing that the c axis is indeed aligned in the direction perpendicular to the glass substrate surface, is also shown. The ceramic involves diffraction peaks from the (100), (002), and (101) planes.

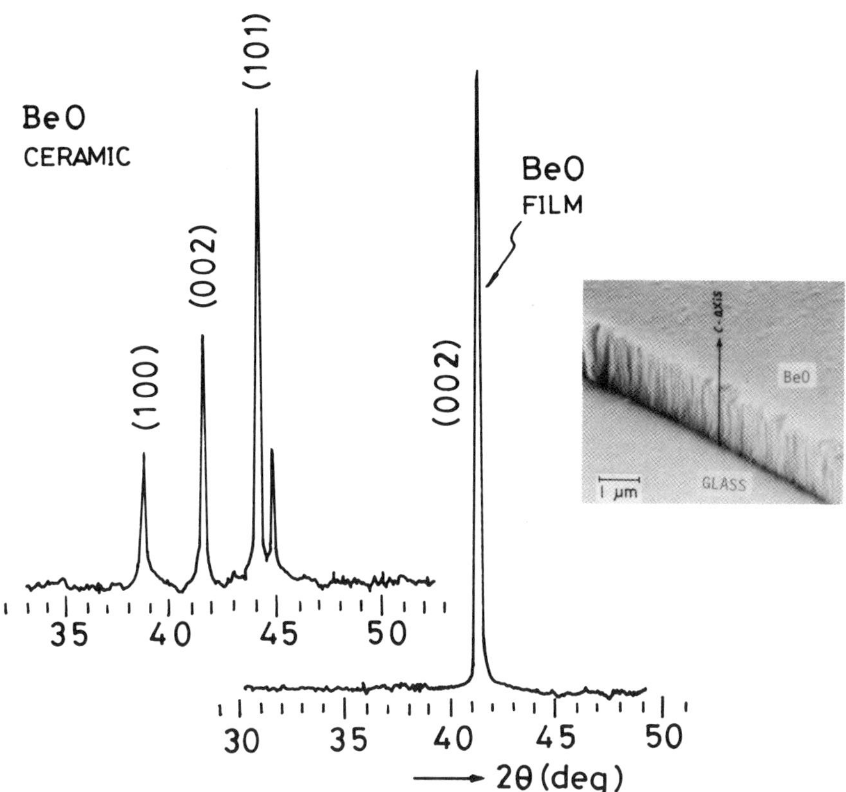

FIG. 45. X-ray diffraction pattern of a BeO film grown on a glass substrate in comparison with that of a BeO ceramic, and the SEM structure of the former.

The film only has the peak from the (002) plane, indicating the *c*-axis orientation.

In order to explain the uniaxially oriented film growth, an *sp* hybridization orbital model for ionized metals (*55*) was used. The results calculated for BeO, ZnO, and AlN are shown in Fig. 46. The potential valleys obtained for ionized metals, e.g., BeO, are predominantly localized along the *r* axis, which corresponds to the *c* axis of hexagonal materials, and their depths are fairly large in comparison with those for neutral species. From the calculation it was found that the potential valley was limited to the *r* axis in this figure, resulting in *c*-axis orientation, and in the case of the neutral species, the corresponding shallow valleys appeared at a value of *r* independent of angle, resulting in isotropic films.

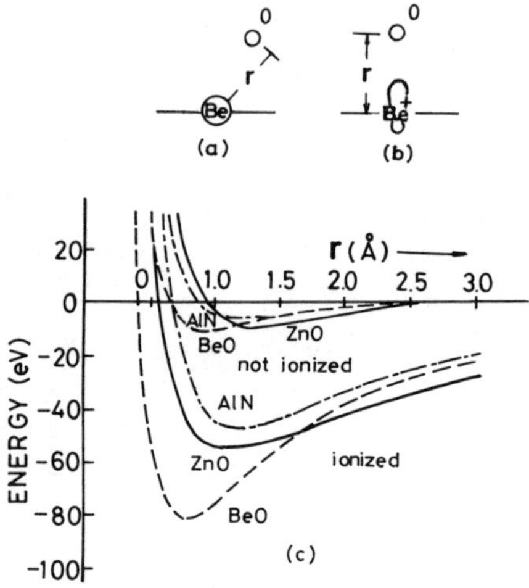

FIG. 46. Model to explain uniaxially oriented BeO film growth. (a) $2s$ orbital for neutral Be atom, (b) sp-hybridized orbital for Be ion, and (c) the energies calculated by varying the distance between the two coordinating atoms (e.g., Be–O), where the r axis corresponds to the c axis of the BeO film.

Measurements of the infrared reflectivities were made in the wavelength region 2.5–25 μm to estimate the dispersion frequency for BeO films (*56*). The experimental results for BeO films grown on sapphire (0001), glass, and Si (111) substrates are shown in Fig. 47 and are compared with those for a pressed plate of BeO that were obtained by Durig, *et al.* (*57*). From the experimental results, the dispersion frequency ω_q^0, which represents the transverse-optical vibration mode, is found to be about 1.39×10^{14}/sec in all films, while ω_q^1, which corresponds to a longitudinal-optical vibration mode, is found to vary with the crystallinity of grown films, and is about 1.79×10^{14}/sec for the sapphire substrate and 2.19×10^{14}/sec for glass and Si substrates.

The value of the static dielectric constant $\varepsilon(0)$ for BeO films was found to be 6.5 (n=2.6) from the value R=0.2 at 400 cm^{-1}. The optical dielectric constant $\varepsilon(\infty)$ was estimated to be 3.29 (n=1.98) for the sapphire substrate and 2.6 (n=1.6) for the glass and Si.

The sound velocity U_s of a phonon could be estimated roughly from the dispersion relation for polar materials. U_s was estimated to be 13.8×10^3 m/sec. This is reasonable as compared with the value 12×10^3 m/sec (*58*), which was measured along the c axis of a flux-grown single crystal.

Anisotropic phonon thermal conductivities were measured with respect to the c-axis-oriented films grown on glass substrates. The thermal con-

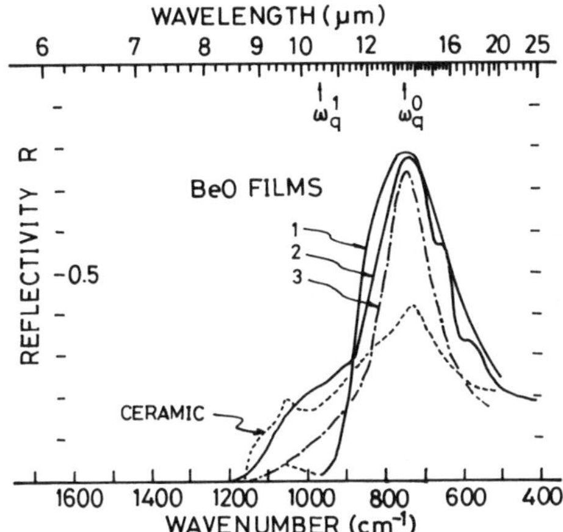

FIG. 47. Infrared reflectivities of BeO films grown on (1) a sapphire (0001) substrate, compared with those on (2) glass and (3) Si(111) substrates.

ductivity $\kappa_{ph}^{\parallel}$, measured with the heat flow parallel to the *c* axis of films, was found to be about four times the value perpendicular to the *c* axis, $\kappa_{ph}^{\perp}$. $\kappa_{ph}^{\parallel}$ at room temperature was typically 2.6 W/cm deg, while $\kappa_{ph}^{\perp}$ was 0.6 W/cm deg. Both $\kappa_{ph}^{\parallel}$ and $\kappa_{ph}^{\perp}$ were found to be proportional to T^{-2} in this temperature region, which can be explained by a simplified model for the thermal conductivity of the lattice applicable for a chemically pure ionic crystal (*59*).

To determine the applicability of this fabrication technology to MOS-type or planar devices, the surface-state density of the interface layer in an Al/BeO/Si structure was examined by means of capacitance voltage measurements at 100 kHz. The experiments suggest that the surface-state density and structural defects were reduced for the *c*-axis-oriented BeO. The surface-state density Q_{ss} obtained in an Al/BeO/Si structure was about 1.6×10^{-8} C/cm^{-2} ($Q_{ss}/q = 1.0 \times 10^{11}$ cm^{-2}), which was lower than that of a Si–SiO_2 native oxide layer ($Q_{ss}/q = 2.5 \times 10^{11}$ cm^{-2}) (*56*).

The *c*-axis-oriented BeO films, therefore, have the potential of developing GHz-band surface-acoustic-wave (SAW) devices, because of (1) high sound velocity, (2) a large electromechanical coupling coefficient to the surface wave, and (3) low propagation loss.

c. FeO_x Films. The reactive ICB technique has been used in fabricating heterostructure-type Si photovoltaic cells coated with iron oxide (*43*). The iron oxide photoelectrode is available as a coating material on semiconductors.

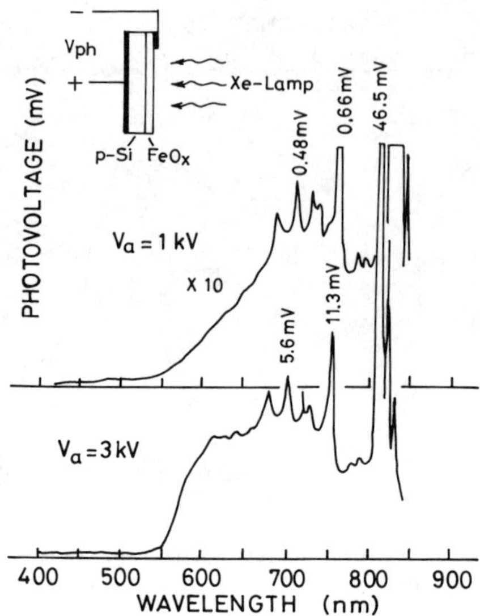

FIG. 48. Photovoltage of n-FeO_X films deposited on a p-Si substrate in the wavelength range 400–850 nm.

The films were prepared by reaction between an Fe-cluster beam and oxygen gas introduced through a leak valve to a pressure of $1\text{–}2 \times 10^{-4}$ torr. p-type Si wafers were used as substrates, and the thickness of the deposited iron oxide films was determined by an interference microscope.

Figure 48 shows the spectral sensitivity of the photovoltage V_{ph} of two heterostructures, in which FeO_x films, about 1000 Å thick, were deposited at $V_a = 1$ and 3 kV. A partially focused 500 W Xe lamp was used as the light source, and the light intensity was about 0.6 W/cm^2. The results indicate the photosensitivity of this type to be strongly enhanced by the acceleration voltage (V_a).

Measurements of the optical absorption coefficient of FeO_x films deposited on glass substrates at different values of V_a were made to study the effect of V_a on the photovoltage. Figure 49 shows the optical bandgap E_g^{opt} calculated from their data. The E_g^{opt} of the film deposited at $V_a = 0$ is about 1.15 eV, which is almost equal to the known value, 1.1 eV, of pure Si. When V_a is 3 kV, E_g^{opt} increases to 1.5 eV, suggesting that the oxygen content of the FeO_x increases with the acceleration voltage.

FeO_x/Si heterostructures of this type have potential advantages for fabricating low-cost solar cells or photoanodic cells which are chemically stable and which have high energy-conversion efficiency.

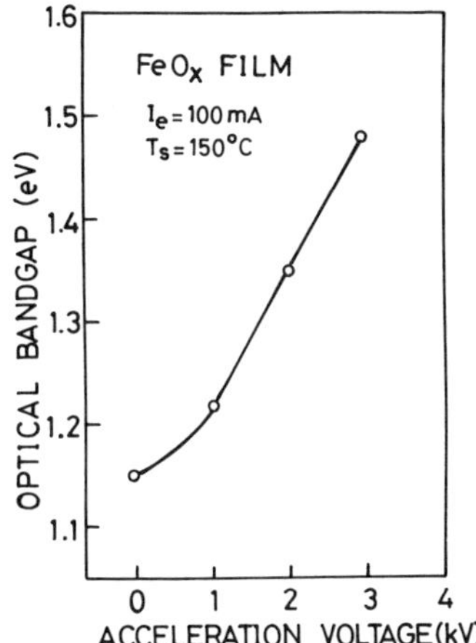

FIG. 49. Dependence of the optical bandgap of FeO_X amorphous films on the acceleration voltage.

d. GaN Films. Usually, GaN LEDs have been made by the epitaxial growth of GaN on sapphire at 1000°C using chemical vapor deposition (CVD) (*60*).

ZnO films are considered favorable as a seed for epitaxial growth of GaN films on amorphous substrates, because ZnO films crystallize easily with the *c* axis perpendicular to the glass substrate, as described previously, and the lattice parameter of ZnO is quite close to that of GaN. The lattice misfit between the (002) planes of GaN and ZnO is only 0.46%, which contrasts with the misfit of 16.4% between the (0001) planes of GaN and sapphire. Epitaxial growth of GaN has been obtained successfully by the RICB technique, even at low substrate temperature (below 600°C) by using a ZnO buffer layer on a glass substrate, which may be termed as "seeded epitaxy" (*44*).

A scanning electron microscope view of the fractured edge of GaN layers grown by RICB at 450°C on the *c*-axis-oriented ZnO films are shown in Fig. 50. The deposition conditions for these GaN/ZnO layers included: electron current for ionization (I_e) of 300 mA, zero acceleration voltage (V_a), and a nitrogen pressure of 5×10^{-4} torr for GaN or the same oxygen pressure for ZnO in the same chamber. The GaN films produced exhibited a resistivity of the order of 500 Ω cm and showed *n*-type conductivity, probably due to a high concentration of the native donor (nitrogen vacancies). The resistivity of the ZnO layer at room temperature was about 1000 Ω cm.

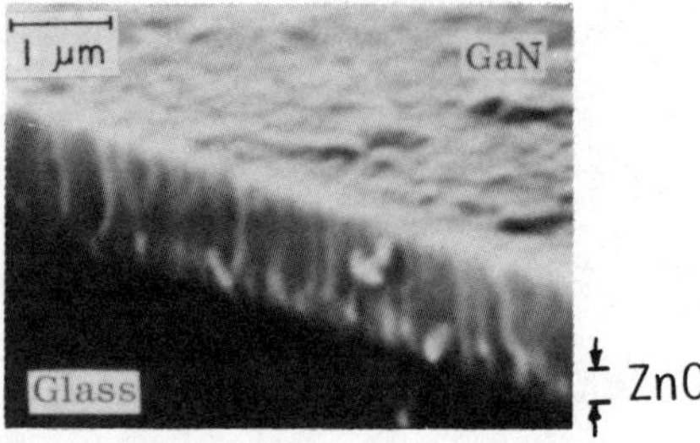

FIG. 50. SEM views of fractured edges of GaN layers on the *c*-axis-oriented ZnO films on a glass substrate.

Measurements of optical absorption coefficients of GaN/ZnO films on a glass substrate were made in the wavelength region from 0.36 to 0.74 μm. The fundamental absorption edge of the grown GaN/ZnO film was determined to be about 3.4 eV. Figure 51 shows the absorption coefficient (α) versus photon energies ($h\nu$) of a grown film, compared with the results of GaN films grown on sapphire substrates by CVD using carrier gases of HCl, NH_3, and H_2 (sample 1) (*60*), $GaBr_3$ and NH_3 (sample 2) (*61*), and by rf sputtering (sample 3) (*62*). The values of a structural parameter E_s of each sample, which were obtained from the slopes of the curves near the band tail, are written in the figure. The E_s for the GaN/ZnO film is about 300 meV, and it is comparable to those of the GaN films grown on sap-

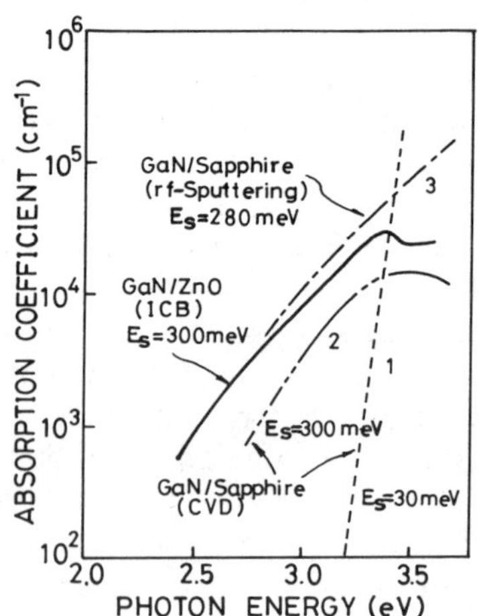

FIG. 51. Absorption coefficient as a function of photon energy for GaN/ZnO films prepared by ICB.

phire substrate except for the value of sample 1 (*60*). A reduction of E_s can be expected by improving the crystallinity of the ZnO layer and the GaN deposited on it, and by adjusting the deposition conditions properly.

e. SiC Films. Silicon carbide films were prepared by RICB deposition (*45*). The vapor of Si ejected through the nozzle reacted with C_2H_2 gas maintained in the 10^{-5}–10^{-4} torr pressure range. Si clusters and C_2H_2 were partially ionized by the electron current (I_e) for ionization of 100–200 mA. The acceleration voltage (V_a) was in the range of 0–8 kV, and the substrate temperature was kept at 400°C.

Figure 52 shows the dependence of the transmittance characteristics of the film on acceleration voltage in the infrared wavelength region. The peak caused by the stretching mode of the Si–C bond is shifted to the high-wavenumber region with increasing acceleration voltage, and the half-width of the peak decreases with increasing acceleration voltage. The peak for the film prepared at V_a = 8 kV appears at a wavenumber of 800 cm^{-1}, which corresponds to the wavenumber due to the TO phonon for crystalline β-SiC. The film obtained is found to be improved in crystallinity with increasing acceleration voltage.

The optical bandgaps measured for the films were in the range of 1.4–1.7 eV, and increased at higher electron currents for ionization and higher acceleration voltages. The increase of the optical bandgap could be due to the ion content and the kinetic energy of the clusters, by which the reaction between Si and C_2H_2 was effectively enhanced.

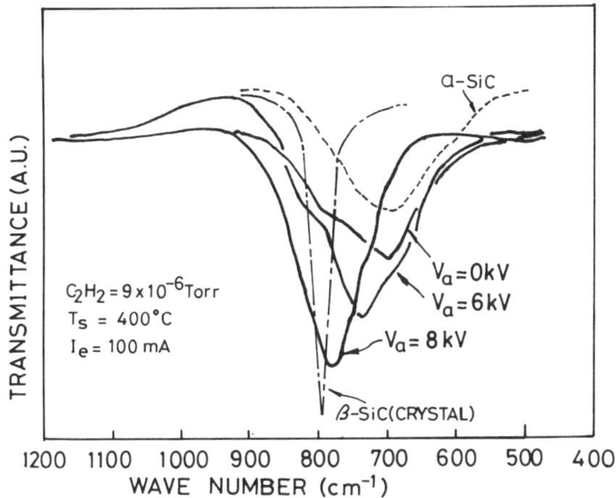

FIG. 52. Transmittance characteristics of SiC films deposited at different acceleration voltages.

6. Organic Material Films

Organic films can be formed by ICB as well as inorganic materials. As one of the examples, anthracene was deposited. The cluster size of anthracene was estimated by energy measurements to be 10–20 molecules. One cluster, therefore, consists of 300–500 carbon and hydrogen atoms. Clusters of anthracene were also formed by homogeneous nucleation in the supersonic nozzle beam during expansion. To confirm the formation mechanism of the anthracene cluster, the following scaling law was used *(63)*

$$P_0 D^q / T_0^r = \text{constant} \tag{10}$$

where $0.5 < q \leqq 1$ and $r = \gamma/(\gamma - 1) - q(2 - \gamma)/(2\gamma - 2)$. The value of q differs with experimental conditions such as the kind of condensing material. Applying the above equations to the experimental result, the optimum value of q was found to be nearly equal to unity. In Fig. 53 the energy of the cluster beam is plotted against $P_0 D^q / T_0^r$, where $q = 1$ and $r = \frac{1}{4}$. All the points corresponding to different nozzle diameters fall on one straight line, which indicates that the formation mechanism of anthracene clusters fits the scaling law. This result shows that the formation of anthracene clusters is due to homogeneous nucleation in the supersonic nozzle beams.

Anthracene films deposited on glass substrates by ICB showed smooth surfaces, strong adhesion, and were highly crystalline. It was found from

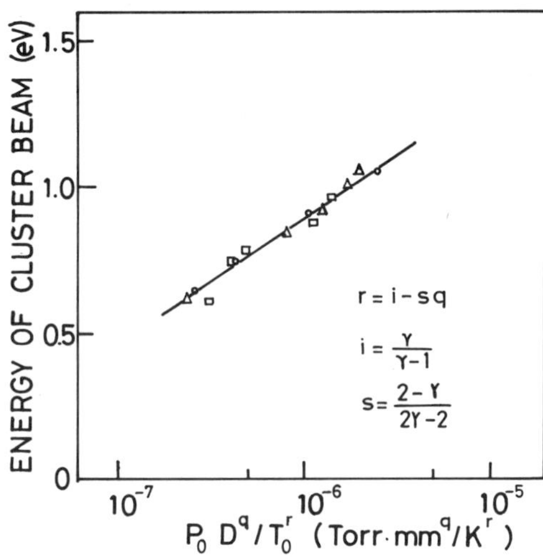

FIG. 53. Application of the scaling law to the formation mechanism of anthracene clusters. Nozzle diameters: (○), 2; (△), 1; and (□), 0.5 mm.

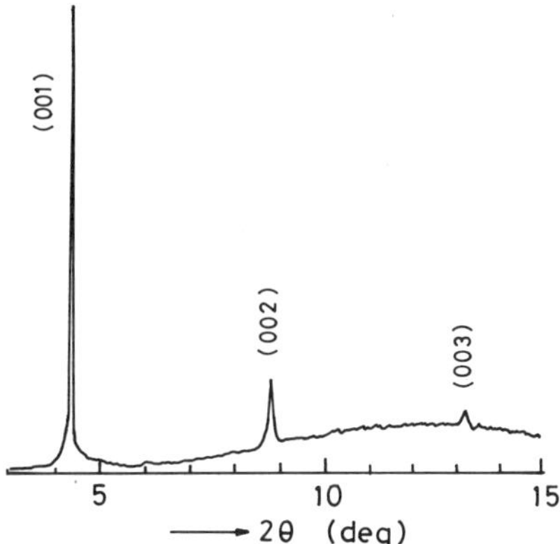

FIG. 54. X-ray diffraction pattern of anthracene thin films prepared by ICB.

the x-ray diffraction pattern of the deposited thin film that the film is preferentially oriented with its (001) plane parallel to the glass substrate, as shown in Fig. 54. The grain size of the film increased by increasing the electron current for ionization to I_e = 10–20 mA, and the acceleration voltage to V_a = 100 V. Also, strong adhesion of the deposited film was obtained, as shown in Fig. 55.

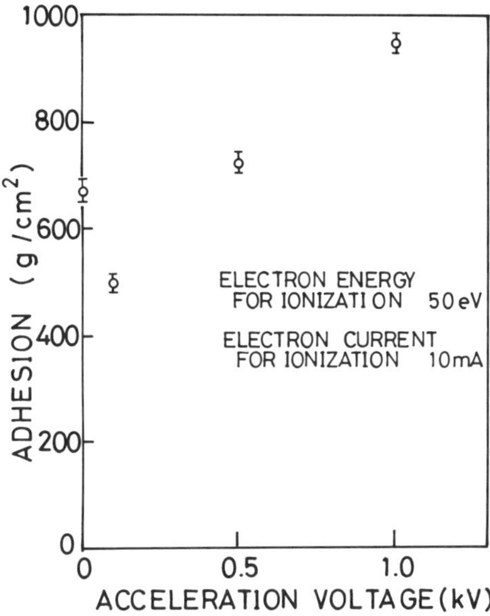

FIG. 55. Dependence of adhesion strength of anthracene films on acceleration voltage.

From photoluminescence measurements, the film prepared at higher acceleration voltage was found to have good crystallinity (*46*). A change of the photoluminescence spectra was seen for the film deposited at a higher acceleration voltage, which showed that some chemical modification was taking place during the film formation. This suggests that the ICB technique may be useful as a new type of polymerization method.

V. Conclusions

The ICB technique allows growth not only of single element or compound material films, but also of hydride, oxide, nitride, and carbide films. The mechanisms of cluster-beam formation, film-formation mechanisms, and some examples of the film properties have been discussed. It has been shown that films could be deposited at lower substrate temperature than those required in some other deposition methods. The remarkable features of ICB versus conventional techniques are due to a unique film-formation mechanism supported by sputtering, heating, implantation, activated-center creation for nuclei, migration effects, and the influence of the presence of charge carried by ionized clusters under equivalently high intense current with suitably low kinetic energy for deposition.

From these experimental results it is clear that the ICB and RICB methods have great potential for the preparation of high-quality coatings ranging from amorphous to monocrystalline for many applications.

References

1. T. Takagi, I. Yamada, M. Kunori, and S. Kobiyama, *Proc. Int. Conf. Ion Sources, 2nd, 1972,* 790, (1973).
2. T. Takagi, K. Matsubara, H. Takaoka, and I. Yamada, *Proc.—Int. Conf. Ion Plat. Allied Tech., 2nd, 1979,* 174 (1979).
3. T. Takagi, *Thin Solid Films* **92,** 1 (1982).
4. C. Weissmantel, *Proc. Int. Vac. Congr., 7th, 1977,* 1533 (1977).
5. E. H. Hirsch and I. K. Varga, *Thin Solid Films* **52,** 445 (1978).
6. T. Takagi, I. Yamada, and A. Sasaki, *J. Vac. Sci. Technol.* **12,** 1128 (1975).
7. T. Takagi, I. Yamada, and A. Sasaki, *Thin Solid Films* **45,** 569 (1977).
8. R. F. C. Fattow, A. G. Cullis, A. J. Grant, G. R. Jones, and R. Clampitt, *Thin Solid Films* **58,** 189 (1979).
9. I. Yamada, H. Takaoka, H. Inokawa, H. Usui, S. C. Cheng, and T. Takagi, *Thin Solid Films* **92,** 137 (1982).
10. T. Takagi, I. Yamada, and A. Sasaki, *Conf. Ser.—Inst. Phys.* **38,** 229 (1978).
11. Technical Data, Eaton Corporation, 16 Tozer Road, Beverly, Massachusetts.

12. Technical Data, Sumitomo Bakelite Co., Totsuka, Yokohama, Japan.
13. R. E. Leckenly, E. J. Robbins, and P. A. Trevalion, *Proc. R. Soc. A London, Ser.* **280,** 409 (1964).
14. O. F. Hagena, *Rarefied Gas Dyn.* **2,** 1465 (1969).
15. P. P. Wegner and G. D. Stein, *Symp. (Int.) Combustion [Proc.]* **12,** 1183 (1969).
16. T. Takagi, "Preprint of Ion Assisted Surface Treatments, Techniques and Processes," 1.1. The Metal Society, London, 1982.
17. J. D. Hirschelder, C. f. Cutiss, and R. B. Bird, "Molecular Theory of Gases and Liquids," p. 336. Wiley, New York, 1964.
18. J. K. Lee, J. A. Barber, and F. F. Abraham, *J. Chem. Phys.* **58,** 3166 (1973).
19. I. Yamada and T. Takagi, *Thin Solid Films* **80,** 105 (1981).
20. J. B Theeten, R. Madar, A. Mircea-Roussel, A. Rocher, and G. Lawrence, *J. Cryst. Growth* **37,** 317 (1979).
21. G. D. Stein, *Phys. Teach.,* Nov., 503 (1979).
22. I. Yamada, G. D. Stein, H. Usui, and T. Takagi, *Proc. Symp. Ion Sources Ion-Assisted Technol., 6th, 1982,* 47 (1982).
23. C. A. Neugebauer, *in* "Handbook of Thin Film Technology" (L. I. Maissel and R. Glang, eds.), Chapter 8. McGraw-Hill, New York, 1970.
24. V. O. Babaev, J. V. Bykov, and M. B. Guseva, *Thin Solid Films* **38,** 1 (1976).
25. K. L. Chopra, "Thin Film Phenomena," Chapter 4. McGraw-Hill, New York, 1969.
26. T. Takagi, I. Yamada, and A. Sasaki, *Thin Solid Films* **39,** 207 (1976).
27. M. Kamiyama and S. Sugata, "Thin Film Handbook," 1-8-4. Ohmusha Syoten, Tokyo, 1964.
28. T. Takagi, K. Matsubara, and H. Takaoka, *J. Appl. Phys.* **51,** 5419 (1980).
29. K. Matsubara, I. Yamada, N. Nagao, K. Tominaga, and T. Takagi, *Surf. Sci.* **86,** 290 (1979).
30. T. Takagi, I. Yamada, and A. Sasaki, *Proc. Int. Vac. Congr., 7th, 1977,* 1915 (1977).
31. T. Takagi, I. Yamada, and A. Sasaki, *Proc.—Int. Conf. Ion Plat. Allied Techn., 1st, 1977,* 50 (1977).
32. T. Ishida, S. Wako, and S. Ushio, *Thin Solid Films* **39,** 227 (1978).
33. H. Inokawa, K. Fukushima, I. Yamada, and T. Takagi, *Proc. Symp. Ion Sources Ion-Assisted Technol., 6th, 1982,* 355 (1982).
34. T. Takagi, K. Matsubara, N. Kondo, K. Fujii, and H. Takaoka, *Jpn. J. Appl. Phys.* **19,** Suppl. 19-1, 107 (1980).
35. N. Kondo, K. Matsubara, and T. Takagi, *J. Magn. Soc Jpn.* **5,** 105 (1981).
36. K. Matsubara, H. Takaoka, K. Shigeno, Y. Kuriyama, and T. Takagi, *Proc. Symp. Ion Sources Ion-Assisted Technol., 6th, 1982,* 399 (1982).
37. T. Koyanagi, K. Matsubara, H. Takaoka, and T. Takagi, *Proc. Symp. Ion Sources Ion-Assisted Technol., 6th, 1982,* 409 (1982).
38. T. Takagi, K. Matsubara, M. Oura, and T. Koyanagi, *Proc. Symp. Ion Sources Ion-Assisted Technol., 6th, 1982,* 391 (1982).
39. I. Yamada, I. Nagai, M. Horie, and T. Takagi, *J. Appl. Phys.* **54,** 1583 (1983).
40. K. Morimoto, H. Watanabe, and S. Itoh, *J. Cryst. Growth* **45,** 334 (1978).
41. T. Takagi, I. Yamada, and A. Sasaki, *in* "Ion Implantation in Semiconductors and Other Materials" (S. Namba ed.), repr., 275. Plenum, New York, 1974.
42. T. Takagi, K. Inoue, S. Mizugaki, A. Sasaki, and I. Yamada, *Vacuum* **22,** 267 (1979).
43. K. Hosono, K. Matsubara, H. Takaoka, and T. Takagi, *Proc. Int'l Ion Eng. Congr.—ISIAT'83 & IPAT'83, Kyoto,* 1237 (1983).
44. K. Matsubara, T. Horibe, H. Takaoka, and T. Takagi, *Proc. Symp. Ion Sources Ion Appl. Technol., 4th, 1980,* 137 (1980).

45. K. Mameno, K. Matsubara and T. Takagi, *Proc. Symp. Ion Sources Ion-Assisted Technol., 6th, 1982,* 341 (1982).
46. H. Usui, M. Naemura, I. Yamada, and T. Takagi, *Proc. Symp. Ion Sources Ion-Assisted Technol., 6th, 1982,* p. 331 (1982)
47. I. Yamada, K. Matsubara, M. Kodama, M. Ozawa, and T. Takagi, *J. Cryst. Growth* **45,** 326 (1978).
48. D. Chen, G. Otto, and F. Schmit, *IEEE Trans. Magn.* **MAG-9,** 66 (1973)
49. N. V. Kolomoets, T. S. Stavitskaia, and L. S. Stilbans, *Sov. Phys.—Tech. Phys.* (*Engl. Transl.*) **2,** 59 (1957).
50. M. Telkes, *J. Appl. Phys.* **25,** 765 (1954).
51. I. Yamada, F. W. Saris, T. Takagi, K. Matsubara, H. Takaoka, and S. Ishiyama, *Jpn. J. Appl. Phys.* **19,** L181 (1980).
52. A. Vecht and N. J. Werring, *J. Appl. Phys.* **3,** 105 (1970).
53. T. Takagi, I. Yamada, A. Sasaki, and T. Ishibashi, *IEEE Trans. Electron Devices* **ED-20,** 1110 (1973).
54. E. Loh, *Solid State Commun.* **2,** 269 (1964).
55. K. Matsubara, Y. Fukumoto, and T. Takagi, *Thin Solid Films* **92,** 65 (1982).
56. K. Matsubara, I. Yamada, H. Takaoka, and T. Takagi, *Jpn. J. Appl. Phys.* **21,** Suppl. 21-1, 403 (1982).
57. J. R. Durig, R. C. Lord, W. J. Gardner, and L. H. Johnston, *J. Opt. Soc. Am.* **52,** 1078 (1962).
58. S. B. Austerman, D. A. Berlincourt, and H. A. Krueger, *J. Appl. Phys.* **34,** (1963).
59. J. Callaway, *Phys. Rev.* **113,** 1046 (1959).
60. J. I. Pankove, H. P. Marusa, and J. E. Berkeyheiser, *Appl. Phys. Lett.* **17,** 197 (1970).
61. J. A. Van Vechten, *Phys. Rev.* **187,** 1007 (1969).
62. Y. Morimoto and S. Ushio, *J. Appl. Phys.* **12,** 1820 (1973).
63. O. F. Hagena, *Phys. Fluids* **17,** 894 (1974).

The Activated Reactive Evaporation Process

R. F. BUNSHAH AND C. DESHPANDEY

Department of Materials Science and Engineering
University of California
Los Angeles, California 90024

I. Introduction

The use of plasma-assisted physical vapor deposition (PAPVD) processes for deposition of compounds (oxides, carbides, nitrides, sulfides, etc.) has spread into various types of industrial applications. These include dielectric films for microelectronics, optical and magnetic applications, hard carbide and nitride films for cutting and forming tools, sulfides for solid-state lubrication and solid electrolytes, etc. In fact, PAPVD methods can be said to have opened up a new area in materials synthesis. Many compound films, which were hitherto difficult to deposit, are now routinely synthesized.

Activated reactive evaporation (ARE) is one of the methods in the general category of plasma-assisted physical vapor deposition processes. It was first developed in 1971 by Bunshah and Raghuram (*1, 2*) at UCLA for the high-rate deposition of refractory compound films using electron-beam evaporation sources. Since then, a considerable amount of work has been carried out by various investigators worldwide on the mechanisms of the process, modifications using an electron-beam evaporation source for low deposition (*3*), modifications for resistance-heated evaporation sources (*4*), the development of variant processes such as low-pressure plasma deposition (LPPD) (*5*) and reactive ion plating (*6*) based on the original ARE process, studies of the relationship between microstructure and process parameters, and the synthesis of coatings for optical, wear, corrosion, and energy-related applications, as well as decorative coatings.

The historical origins of reactive evaporation processes go back to 1907, when Soddy (*7*) found that calcium vapor reacted with gases except the inert gases. In 1913 Langmuir (*8*) studied the formation of tungsten nitride by vapor-phase reaction between tungsten and nitrogen. The use of reactive evaporation processes in the deposition of oxide films goes back to the pioneering work of Brinsmaid *et al.* (*9*) in 1957 and Auwarter (*10*) in 1960, who studied the deposition of oxide films by reaction between metal or suboxide vapors and oxygen gas. Auwarter (*10*) also suggested that reactivity can be enhanced by separate ionization of the oxygen gas molecules using an electrical discharge, prior to interaction with the metal atoms.

Other work on reactive evaporation processes without ionization of the reactive gas is as follows:

(1) Herrick and Tevebaugh (*11*) deposited copper oxide films by vaporization of copper from resistance-heated sources in an oxygen atmosphere.

(2) Novice *et al.* (*12*) and Schilling (*13*) deposited Al_2O_3 films by reactive evaporation from a resistance-heated aluminum source in the presence of oxygen.

(3) Ritter (*14*) produced thin films of Si_2O_3 and TiO_2 by reactive evaporation of Si, Ti, SiO, and TiO from resistance-heated sources in the presence of 10^{-4} to 10^{-3} torr partial pressure of oxygen.

(4) Ferrieu and Pruniaux (*15*) produced Al_2O_3 by reactive evaporation of Al in an atmosphere of water vapor at 10^{-1} torr in the reaction zone.

(5) Rairden (*16*) prepared thin films of NbN and TaN by evaporation of Nb and Ta from an electron-beam-heated source in an N_2 partial pressure of 10^{-4} to 10^{-3} torr, and AlN by evaporation of Al in an NH_3 atmosphere (*17*).

(6) DeKlerk and Kelly (*18*) produced CdS and ZnS films by coevaporation from two independent sources of Cd/Zn and S and condensation on a substrate at temperatures between 50 and 200°C.

(7) Learn and Haq (*19*) produced β-SiC by reactive evaporation of Si in a C_2H_2 atmosphere.

Examples of reactive evaporation processes where the reactive gas was ionized in a separate chamber (located inside or outside the vacuum system) are as follows:

(1) Auwarter (*10*) studied the deposition of thin films oxides of Si, Zr, Ti, Al, Zn, and Sn by reactive evaporation of the metal from resistance-heated sources in a partial pressure of oxygen gas. Ionization of the oxygen gas outside the reaction zone by glow discharge between two electrodes is claimed to increase the "affinity" between the gas ion and the metal compound, i.e., to enhance the probability of formation of metal compounds. Deposition rates of about 0.2 μm/min were obtained.

(2) Wank and Winslow (*20*) deposited films of AlN by evaporating films of Al from an rf-heated BN crucible and reacting the Al deposited on the substrate with N_2 gas which has been dissociated by 60 Hz ac discharge at the end of the gas feed tube. Deposition rates of 0.2 to 0.2 μm/min were obtained.

(3) Kosicki and Khang (*21*) produced GaN thin films by depositing pure Ga from resistance-heated sources onto a substrate in the presence of activated N_2 gas. The N_2 gas was made chemically active by partial dissociation in a microwave discharge located away from the source and the substrate. Deposition rates of 0.2 and 0.3 μm/min were obtained.

(4) Heitmann (*22, 23*) used a hollow cathode discharge in a glass chamber to ionize oxygen gas and deposit films of SiO_2, SiO_xN_y, and TiO_2.

(5) More recently, Küster and Ebert (*24*) have used a modification of the method of by Heitmann to deposit TiO_2 layers and study their optical

properties. These authors refer to this process as the activated reactive evaporation process.

(6) A later contribution is an excellent and detailed paper by Ebert (*24a*) on the deposition of TiO_2, BeO, In_2O_3, SnO_2, and SiO_2 coatings using ionized oxygen gas.

At this time it should be pointed out that there is an important distinction between the reactive evaporation processes using an ionized gas stream (as detailed above) and the activated reactive evaporation process (or other plasma-assisted deposition processes). In the ARE process, ionization of *both the metal vapor and the reactive gas or gas mixture* occurs in the reaction zone, which is defined as the space between the metal vapor source and the substrate.

II. Processes for the Deposition of Refractory Compounds

Refractory compounds are substances like oxides, nitrides, borides, and sulfides that characteristically have a very high melting point (with some exceptions). In some cases, they form extensive defect structures, i.e., exist over a wide stoichiometric range. For example, in TiC, the [C]/[Ti] ratio can vary from 0.5 to 1.0, demonstrating vacant carbon lattice sites. In other compounds, the stoichiometric range is not so wide.

Evaporation processes for the deposition of refractory compounds are further subdivided into three types: (1) direct evaporation (*25*), where the evaporant is the refractory compound itself; (2) reactive evaporation (*9, 10*), where a metal or low-valence compound is evaporated in the presence of a partial pressure of a reactive gas to form a compound deposit, e.g., where Ti is evaporated in the presence of a partial pressure of a reactive gas to form a compound deposit, where Ti is evaporated in the presence of N_2 to form TiN, or where Si or SiO is evaporated in the presence of O_2 to form SiO_2; and (3) activated reactive evaporation (*1*), where metal is evaporated in the presence of the plasma of the reactive gas, e.g., Ti in C_2H_2 plasma to form TiC.

III. Direct Evaporation

Evaporation can occur with or without dissociation of the compound into fragments. The observed vapor species show that very few compounds evaporate without dissociation. Examples are SiO, MgF_2, B_2O_3, CaF_2, and other Group-IV divalent oxides (SiO homologs like GeO and SnO).

In the more general case, when a compound is evaporated or sputtered, the material is not transformed to the vapor state as compound molecules but as fragments thereof. This compound fragmentation step is very difficult to characterize and control. Subsequently, the fragments have to recombine, most probably on the substrate to reconstitute the compound. Therefore, the stoichiometry (anion-to-cation ratio) of the deposit depends on several factors, including the deposition rate and the ratios of the various molecular fragments, the impingement rate of other gases present in the environment, the surface mobility of the fragments (which in turn depends on their kinetic energy and substrate temperature), the mean residence time of the fragments on the substrate, the reaction rate of the fragments on the substrate to reconstitute the compound, and the impurities present on the substrate. For example, it was found that direct evaporation of Al_2O_3 resulted in a deposit which was deficient in oxygen, i.e., which had the composition Al_2O_{3-x}. This O_2 deficiency could be made up by introducing O_2 at a low partial pressure into the environment (*26*).

In other cases, the situation is more complex (*27*). ZrB_2 deposits produced by direct evaporation of ZrB_2 billets from an electron-beam-heated source either at high or low deposition rates (2.14 and 0.11 μm thickness per minute, respectively) consisted entirely of the ZrB_2 phase. On the other hand, for similar experiments with high-rate evaporation of TiB_2, the deposits consisted of a mixture of TiB_2 and TiB phases with the amount of the TiB phase increasing with higher deposition temperatures. Low-rate evaporation of TiB_2 produced TiB_2 deposits exclusively.

A further operational problem occurs due to the physical disintegration by cracking of the evaporant billet under the impact of the electron beam due to the poor thermal conductivity of the refractory compound materials and the method of billet fabrication, i.e., pressing and sintering of refractory compound powders, which results in incorporation of gas pockets in the compound that expand on heating and cause high stresses, leading to crack formation and propagation. This is not an insuperable problem, as direct evaporation, especially at low rates, can be carried out from a source consisting of coarse powder or pebble material. In some cases, fused refractory compounds such as fused alumina or silica are available and serve as good evaporation materials.

IV. Reactive Evaporation Processes

The difficulties involved due to fragmentation of the compounds in direct evaporation processes are overcome in reactive evaporation, where the metal is evaporated in the presence of the reactive gas. The compound is formed by reaction of the evaporating metal species with the molecules of the reactive gas. Although this technique has been extensively used to deposit a variety of oxide films for optical applications, it is generally observed that the films are slightly deficient in oxygen. Moreover, it is also observed that in some cases, especially in the synthesis of carbide films, the deposition rate becomes a limiting factor governing the growth of the films. In such cases stoichiometric TiC films could only be deposited at very low rates (~1.5 Å/sec maximum) (*28*). This limitation of deposition rate in the case of reactive evaporation is due to the reaction kinetics involved in the formation of the compound by this process. The presence of "plasma" in the ARE process influences the reaction kinetics by providing activation energy to the reactive species, thereby making it possible to synthesize compound films at considerably higher rates. The role of the plasma in activated reactive evaporation is discussed below. A brief comparison between the activated reactive evaporation and reactive sputtering techniques is included in the discussion to elucidate the differences between the two.

V. Activated Reactive Evaporation and the Role of Plasma

The ARE process generally involves evaporation of a metal or an alloy in the presence of the plasma of reactive gas (*2, 29*). For example, TiC and TiN coatings are deposited by this process by evaporating Ti in the presence of C_2H_2 and N_2 plasma, respectively. The major role of plasma in this process is twofold.

(1) To enhance the reactions that are necessary for the deposition of compound films.
(2) To modify the growth kinetics and hence the structure/morphology of the deposits.

In the following sections we will discuss the above two aspects with particular reference to carbide and nitride coatings.

1. Thermodynamic and Kinetic Effects of Plasma

In order to understand the role of plasma in enhancing the chemical reactions essential for the formation of a particular compound, one has to consider the kinetics of these reactions. For the formation of a compound by any chemical reaction, the corresponding thermodynamic and kinetic constraints must be satisfied. Thus, the thermodynamic and kinetic constraints also apply to the deposition of refractory compound films by reactive evaporation.

Let us consider the reactions involved in the synthesis of some oxides, carbides, and nitrides by reactive evaporation in view of the discussion given above. Given below are the reactions for forming Al_2O_3, TiC, and TiN.

$$2Al + \tfrac{3}{2}O_2 \rightarrow Al_2O_3 \qquad \Delta G^\circ = -250 \text{ kcal (mol } O_2)^{-1} \text{ at 298 K}$$

$$2Ti + C_2H_2 \rightarrow 2TiC + H_2 \qquad \Delta G^\circ = -76.5 \text{ kcal(mol } C_2H_2)^{-1} \text{ at 298 K}$$

$$2Ti + N_2 \rightarrow 2TiN \qquad \Delta G^\circ = -73.5 \text{ kcal (mol } N_2)^{-1} \text{ at 298 K}$$

As can be seen from the above reactions, the thermodynamic criterion of negative free energy of formation is satisfied for the respective compounds.

The reaction kinetics in reactive evaporation process can be treated in exactly the same manner as for reactions occurring in heterogeneous systems of condensed phases. The model for heterogeneous metallurgical reactions involves: (1) transport of reactant to the reaction interface, (2) transport of reaction products away from the reaction interface, (3) the chemical reaction at the chemical interface, (4) the nucleation of new phase, and (5) heat transfer to or away from the reaction interface.

For reactive evaporation this model may be depicted as follows (e.g., for TiC formation):

Reactants	Products
Ti (metal atoms)	TiC (deposit)
C_2H_2 (gas)	H_2 (gas)

Reaction
interface

On the basis of the above model, the rate-controlling steps in the reactive evaporation process are (1) adequate supply of reactants, (2) adequate collision frequency, (3) the rate of chemical reactions at the interface, and (4) the rate of removal of the reaction products from the interface.

It is easy to satisfy (1), (2), and (4) above for a reactive evaporation process. However, condition (3), i.e., rate of reaction, becomes the rate-governing step. *The "plasma" in the ARE process influences this step, i.e., the rate of reaction, by providing the necessary activation energy to the reactive species.* This effect of plasma on rate of reaction can be clearly demonstrated by considering the results of Abe *et al.* (*28*) and Bunshah and Raghuram (*1*) on deposition of TiC coatings. Abe *et al.* found that titanium carbide with a carbon-to-titanium ratio of 1 could be formed by a reaction between Ti and C_2H_2 or C_2H_4 molecules on the substrate at 300–500°C only if the deposition rate was ≤1.5 Å/sec. At higher rates from 2 to 4 Å/sec the ratio of carbon to titanium decreased from 1 to 0.02. Clearly the activation barrier could not be overcome at the higher deposition rates. Bunshah and Raghuram (*1, 30*) have similarly reported that the deposition of TiC at high rates required a very high substrate temperature, exceeding 1000°C. However, in the presence of plasma, these authors reported that it was possible to deposit TiC at a considerable rate at a relatively low substrate temperature. The plasma imparts sufficient energy to the reacting species to overcome the activation barrier, and hence condition (3), i.e., the rate of reaction, no longer remains the rate-governing step.

2. Influence of Plasma on the Growth Kinetics of the Deposits

In order to understand the role of plasma on the overall growth kinetics of the depositing film, one has to consider its influence on the three characteristic steps involved in the formation of the deposit, viz., (1) creation of the vapor phase, (2) transport of the vapor phase, and (3) film growth on the substrate. The above three steps are shown in Fig. 1. In the following discussion, we will comment on the effect of plasma on the above three steps with particular reference to the activated reactive evaporation process. A brief comparison with reactive sputtering will be given to elucidate the differences that exist between the two techniques.

Fig. 1. The three steps in film deposition.

a. Plasma–Source Reactions. In the ARE process, the vapor species are generated by the thermal energy imparted to the target. The evaporation rate varies directly as the vapor pressure of the target element, which in turn is dependent on the temperature of the target surface. The plasma therefore has little or no influence on the evaporation rate. Therefore the vapor generation rate in the ARE process is plasma independent.

Unlike ARE, however, in the reactive sputtering process the vapor species are generated by momentum exchange between the positive ions bombarding the target and the atoms of the target. The sputtering rate is therefore totally dependent on the power input to the target, i.e,, the cathode voltage and current for dc and rf sputtering. Thus the sputtering rate in this case is plasma dependent. However, this is not true in the case of the ion beam sputtering process. Moreover, the plasma–target interaction in reactive sputtering gives rise to target poisoning effects (*31*). The target poisoning makes it difficult to achieve high deposition rates for most of the compounds by reactive sputtering techniques. In recent years, many different approaches have been proposed to get around the problem of target poisoning (*32–37*). However, most of these solutions seem to work under limited conditions, and it is difficult to obtain generalized solutions. For a detailed discussion on this subject, the reader is referred to a recent review by Deshpandey and Bunshah (*38*).

In contrast to reactive sputtering, plasma–source interactions do not pose the serious problem of limiting the growth rate of compound films when deposited by activated reactive evaporation. Any compound formed on the surface of the evaporating metal is likely to be dissociated due to the high temperature of the source, particularly under the impact spot of the electron beam or cathodic arc. Also the stirring action in the liquid pool tends to push away any compound layer, thus providing a metallic surface at the top of the liquid pool. It is owing to these differences in plasma–source interactions that it is possible to deposit stoichiometric compounds, vary their stoichiometry, deposit intermediate phases, deposit two-phase mixtures, etc. by the ARE technique.

b. Plasma Interactions During Transport. Numerous reactions can take place due to plasma–vapor interactions during transport of material from the source to the substrate (plasma volume reactions). The reactions that are of importance in reactive PAPVD processes are electron-impact excitation, ionization, and dissociation. For example

$$e^- + x \rightarrow x^* + e^-, \qquad e^- + x \rightarrow x^+ + 2e^-, \qquad e^- + x \rightarrow A^+ + B + 2e^-$$

The rates of these reactions can be represented (*39*) as

$$R = n_e K_i [x] \tag{1}$$

where n_e is electron concentration, K_i is the rate constant, and [x] is the concentration of x. The rate constant can be represented as follows (*40*)

$$K_i = \left(\frac{2}{m_e}\right)^{1/2} \int Ef(E)\sigma_i(E)\, dE \tag{2}$$

where m_e is the electronic mass, $f(E)$ the electron distribution function, and $\sigma_i(E)$ the collision cross section for the particular reaction. Thus, using Eqs. (1) and (2), one can estimate the rate of formation of particular species in the glow discharge. Thornton (*41*) has discussed the analytical model illustrating the principal mechanism of radical formation in glow discharges. A detailed calculation is reported by Kushner (*42*). A variety of radicals, metastable species as well as excited and ionized species, are generated in the plasma by a combination of the reactions cited above.

Numerous studies on the preparation of *a*-Si by glow discharge techniques (*40, 43–45*), as well as studies on plasma etching (*46*) and plasma polymerization (*47*), can be cited to illustrate the importance of plasma volume interactions. More recently, Aita and Myers (*48*) have demonstrated how a change in Ta^+ flux on the substrate can influence the TaN film properties. Ta^+ ions were formed in the plasma during transport by Penning ionization. Such detailed studies on the ARE technique have not been carried out. Similarly, plasma volume interactions in ARE would also be expected to play a dominant role in governing properties of the film. As discussed by Yee (*49*), the chemical characteristics and interactions within the plasma zone can have significant effects on the properties of the final coating deposited by the ARE technique. The above-mentioned arguments can be substantiated by considering the example of TiC coatings prepared by ARE. Bunshah and Raghuram (*1, 30*) found that when Ti was evaporated in a partial pressure of C_2H_2, TiC coatings were formed only in the presence of a plasma, whereas in the absence of activation (i.e., no plasma) the deposit contained only Ti and C. This result was confirmed much later by Abe *et al.* (*28*), who found that in the absence of plasma, TiC with a [C/Ti] ratio of 1 could only be obtained at very low deposition rates of 1.5 Å/sec. At higher rates of 2–4 Å/sec, the [C/Ti] ratio decreased to 0.2, and the films contained large amounts of free Ti and C. Thus, in the absence of the plasma, the activation energy barrier for the reaction between Ti and C_2H_2 species is difficult to surmount (*50*). Yee (*49*) has proposed a model for the synthesis of TiC by the ARE process from Ti and C_2H_2 gas. The following discussions reflect the content of Yee's paper. He considers the ARE process as one in which the substrate is located outside but near the plasma zone, so that the fate of the activated species must be considered as they move toward the sub-

TABLE I

EMISSION SPECTRA OF GASEOUS ATOMS AND MOLECULES OBSERVED IN THE MICROWAVE DISCHARGE OF SOME CHEMICAL SYSTEMS. THE SPECTRAL RANGE IS 2000–6800 Å (*49*)

Chemical system	Species observed
N_2/Se(solid)	SeN, N_2, Se_2, Se
$PCl_3/AsCl_3$/Ar	AsP, P_2, As_2, P, As, Cl, Cl_2^+
$PCl_3/SbCl_3$/Ar	PSb, P_2, P, Sb, Cl, Cl_2^+
$AsCl_3/SbCl_3$/Ar	AsSb, As_2, As, Sb, Cl, Cl_2^+
C_2H_2, C_2H_2/Ar, C_2H_2/He	CH, CH^+, C_2, C
CH_4, CH_4/Ar	CH, CH^+, C_2, C
$TiCl_4$/Ar(or He)	TiCl, Cl_2^+, Ti, Cl
$TiCl_4/N_2$	TiCl, N_2, Cl_2^+, Ti, Cl
$TiCl_4/N_2$/He	TiN, TiCl, N_2, Cl_2^+, Ti, Cl
$TiCl_4/C_2H_2$/He	TiCl, CH, CH^+, C_2, HCl^+, Cl_2^+, Ti, C, Cl
$TiCl_4/CCl_4$	TiCl, CCl, C_2, Cl_2^+, Ti, C, Cl

strate surface. If these species are known, then the actual elementary chemical reactions occurring at the substrate can be considered. The transient species relevant to this paper are given in Table I and are determined by emission spectroscopy in the range of 2000–6800 Å. The absolute or relative concentrations are not known, and in general their determination is very difficult. The relative concentrations of ions is generally quite small due to their high excitation energies. In a given system, the relative concentrations of the species depend on the experimental parameters such as pressure, electric field strength, and the type of discharge carrier gas.

Let us now consider the synthesis of TiN. In the absence of activation, i.e., reactive evaporation, the reaction is

$$2\mathrm{Ti}(^3F_2) + \mathrm{N}_2(X\ ^1\Sigma_g^+) \rightarrow 2\mathrm{TiN} \tag{3}$$

where the reactants are in their ground electronic states. This reaction is exothermic at room temperature so that, in principle, TiN can be formed in the gas phase or on the substrate surface. With activation, the formation of TiN outside the plasma zone is no longer limited to the reaction of Eq. (3) since low-lying metastable atoms of Ti and N exist. In addition, there is the possibility of vibrationally excited N_2 in the ground state such that there is a vibrational thermal disequilibrium. As a result of activation, five exothermic reactions outside but near the plasma zone are possible.

These are

$$Ti(^3F_2) + N(^4S^0) + M \rightarrow TiN \tag{4}$$

$$Ti(^3F_2) + N^*(^2D^0, {}^2P^0) + M \rightarrow TiN + M \tag{5}$$

$$2Ti^*(^5F, {}^1D, {}^3P) + N_2(X\ {}^1\Sigma_g^+, v \Rightarrow 0) \rightarrow TiN \tag{6}$$

$$Ti^*(^5F, {}^1D, {}^3P) + N(^4S^0 + M \rightarrow TiN + M \tag{7}$$

$$Ti^*(^5F, {}^1D, {}^3P) + N^*(^2D^0, {}^2P^0) + M \rightarrow TiN + M \tag{8}$$

The relative amounts of N atoms are small, hence the reactions in Eqs. (3) and (6) should be the dominating reactions in the coating process. Inside the plasma zone, the number of reactions which may lead to the formation of TiN is much greater because of the presence of highly excited species. In addition, if the Ti_2 molecule is present in the plasma phase, then the elementary reactions are much more complicated.

The synthesis of TiC coatings by reaction between Ti and C_2H_2 is also of considerable interest. The plasma zone of this system undoubtedly contains the electronically excited species CH, CH^+, and C_2 as well as Ti, Ti^+, H_2, and H. Outside the plasma zone, but near it, the important species are Ti, HC, C_2, and C, mostly in their ground electronic states. In this case, there are three possible exothermic reactions that may lead to the formation of gaseous TiC. They are

$$Ti + C + M \rightarrow TiC + M \tag{9}$$

$$Ti + C_2 \rightarrow 2TiC \tag{10}$$

$$Ti + CH \rightarrow TiC + H \tag{11}$$

If these species are also produced in their low-lying metastable states, then there are all together 12 exothermic reactions. However, the reactions involving ground-state species are likely to be more important. This is more so as the distance between the plasma zone and the substrate surface increases. Regardless of their relative contributions, all of these reactions are exothermic, and all involve the transient species CH, C_2, and C which are produced in the plasma zone. If the substrate is located inside the plasma zone, the reactions are much more complicated because of the more highly excited species. The reactions that may dominate would depend on the relative concentration and lifetime of the above species, which are strongly affected by process parameters. It is interesting to consider at this point the results of Stowell (*51*) on ion plating of TiC, which substantiates the importance of plasma volume reactions. He observed that codeposits of Ti and carbon were formed on the substrate when attempts were made to deposit TiC at C_2H_4 pressures in the 1–20 μm range typically used in "ion plating." This is likely due to (1) unavailability of Ti in excited/ionized form and (2) reaction of carbon–carbon species to form carbon in the gas phase which can then scatter deposit on the substrate surface. (This observation is analogous to TiC deposition by reactive evaporation, i.e., without a plasma.)

Fukutomi *et al.* (*52*) have reported a comparative study of the properties of TiC coatings prepared using dc and rf excitation. They observed that it is easier to obtain stoichiometric TiC when a dc discharge is used, whereas carbon-rich TiC is likely to be formed in rf excited discharge due to the dissociation of the hydrocarbon gas by the more energetic electrons in the rf discharge. The preferred orientation of the films deposited using dc and rf excitation, as shown in Fig. 2, was also found to be different by

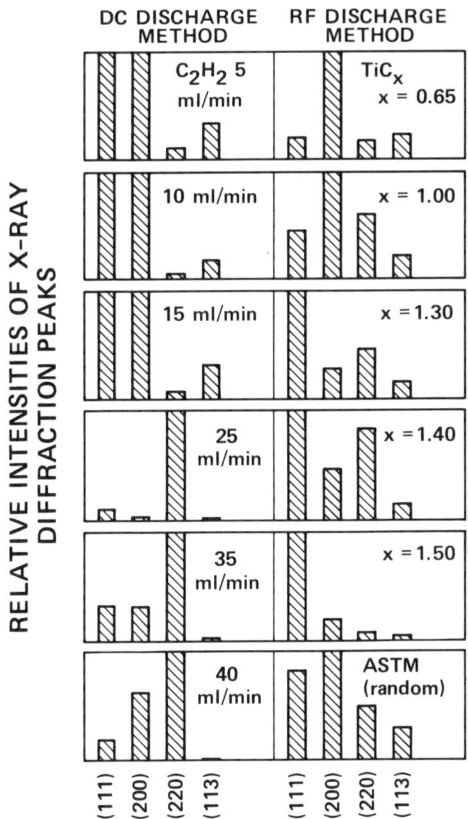

FIG. 2. Dependence of preferred orientation of TiC deposits on discharge conditions (*52*).

these authors. Based on specimen current characteristics, these authors concluded that the differences in stoichiometry and structure are due to differences in ionization and excitation reactions that take place in dc and rf discharges.

The above discussion elucidates the importance of plasma volume reactions (and consequently the radical and species generation) on the growth mechanism of compound films prepared by reactive PAPVD processes. Most of these reactions are dependent on electron energy and electron distribution function (*31, 40, 41*). Since these parameters are different in reactive sputtering and activated reactive evaporation, it is likely that the plasma volume reactions may be different in these techniques.

Even though one can assume that the electron distribution function is Maxwellian for both the ARE and reactive sputtering processes, it is likely that the average electron energy and electron density could be different. As shown by Chapman (*53*), a change in the average energy of electrons can alter the electron fraction in a given energy range significantly. The rate of the various plasma volume reactions may therefore differ significantly in the respective techniques, as can be seen from Eqs. (1) and (2) above. This would lead to differences in concentrations of various species in the region between the source and the substrate in sputtering and ARE processes. Hence the plasma chemistry in the respective techniques will be significantly different, as it is dependent on the relative concentration (and energy) of the species present.

Since the plasma volume reactions are dependent on electron energy and distribution function, it would be advantageous if one can control them independently of other deposition parameters such as pressure, input power, etc. Such an independent control of electron energy and distribution function can be achieved in the ARE process, whereas in reactive sputtering it is not possible since electron energy, pressure, and target voltage are interdependent. The above discussion is aimed at illustrating possible differences in the two techniques. Unfortunately, no direct detailed studies on the ARE plasma have been reported to substantiate the arguments cited above.

The above discussion illustrates the complexity of the problem relative to the deposition of compound films by plasma-assisted deposition processes. To sort out the dominating or rate-controlling reactions is a formidable experimental problem which has yet to be tackled! At present, one can consider a simple-minded concept; i.e., the deposition of compound films involves reactions in the plasma leading to the formation of intermediate reaction products which subsequently react (most probably) on the substrate to produce the compound film. However, reaction of these intermediate products in the gas phase particularly at or in the boundary layer next to the substrate cannot be ruled out.

c. Plasma–Substrate Interactions. Substrates exposed to a glow discharge are bombarded by energetic neutrals, ions, and electrons. The nature and energy of the bombarding species are primarily dependent on the process parameters and geometrical location of the substrate within and/or outside the plasma zone (*54, 55*). Such bombardment can initiate a variety of reactions which may lead to substrate heating, substrate surface chemistry changes, re-emission or sputtering of deposited material, gas incorporation in the growing film, as well as modification of the film morphology, crystal orientation, grain size, etc. Thus, substrate bombardment can have a pronounced effect on the properties of the film.

Substrate bombardment in a glow discharge is a consequence of the potential developed on the surface with respect to the plasma. Because of the difference in mobility of electrons and ions, a space-charge region (sheath) forms adjacent to the surface in contact with the plasma from which one species is largely excluded. The nature of the sheath depends on the current density passing across it. The function of the sheath is to produce a potential barrier so that the more mobile species, i.e., electrons, are electrostatically deflected away from the substrate. The height of the potential barrier thus adjusts itself so as to balance the electron flow to the surface to be equal to that drawn out in the external circuit. Thus, any surface in contact with a plasma develops a potential which is slightly negative with respect to the plasma. The potential is referred to as the floating potential. Substrate bombardment is dependent on this potential (as energetic ions are accelerated across this potential), which in turn depends on the electron energy and distribution function (*31, 41*). In order to control the substrate reactions and thereby modify the film growth, one has to be able to control the electron energy and distribution function independently of other process parameters. It is in this respect that differences exist between diode-type reactive sputtering and activated reactive evaporation, as discussed below.

In conventional diode-type reactive sputtering using dc or rf, the floating potential is largely dependent on pressure and power, i.e., the target potential. These parameters, on the other hand, also govern the deposition rate. It is this interdependence of electron energy on other operating parameters which makes it difficult to obtain an independent control on substrate potential and the bombardment of the growing film. The situation becomes complex, especially in case of deposition of dielectric films such as Al_2O_3, in contrast to the deposition of conducting films (*54, 55*). Intuitive solutions such as locating substrates well outside the plasma region are not practically viable since the deposition rate is drastically reduced with an increase in target–substrate distance. Biasing the substrates is another alternative. However, it must be noted that the substrate bombardment depends on its surface potential with respect to the plasma, which is given by $V_B + V_P$, where V_B is the applied bias and V_P is the plasma potential with respect to ground. It should be noted that V_P also depends on target potential and pressure. Whether this would have a significant effect on film properties would depend on the actual value of V_P in comparison to V_B. Hence, in order to isolate the effect of substrate bombardment on film properties, it is necessary to measure V_P, which can vary significantly under different conditions (*56*). In the case of magnetron sputtering, however, the substrate potential can be altered independently, since the substrates in this case are well outside the plasma zone.

The above discussion holds good for ARE processes for the case where the substrate is located within the plasma. However, in contrast to sputtering, the source–substrate distances can be increased in ARE without a significant penalty in deposition rate since the rate is solely dependent on source temperature (for a given pressure and substrate distance), which is a parameter independent of the plasma and can be increased to compensate for the longer source–substrate distance. For the case where the substrate is essentially out of the plasma zone, its floating potential tends to be a negligibly small value which cannot have any significant effect on V_B.

Since the substrate bombardment is different in all three processes, it is difficult to make clear comparisons of the properties of the films deposited by these processes. Also, it is difficult to estimate how critically these differences influence the properties of the coatings, since no comparative studies have been reported.

It is interesting to note the results of investigations on biased sputtering of Ti in Ar + N_2 plasma, as reported by Poitevin and Lemperiere (*57*). They observe that with increasing substrate bias the ionization increases because many secondary electrons emitted from the cathode return to the negative glow due to reflection from the negative ion sheath. As a result, the discharge becomes more conducting, and for constant power the discharge voltage decreases, causing a reduction in deposition rate. Although these results are useful in illustrating the role of substrate bias in the ionization processes occurring in the glow discharge, no film properties have been reported by the authors. This lack of data makes it difficult to comment further on plasma–film property correlations.

VI. Implementation of the Activated Reactive Evaporation Process

1. Basic Variants of the ARE Process

The two basic variants of the ARE process are activated reactive evaporation with an electron-beam evaporation source (*2*) and the ARE process with a resistance-heated source (*58*).

a. ARE Processes with an Electron-Beam-Heated Evaporation Source. This is illustrated in Fig. 3. In this process, the metal is heated and melted by a high-acceleration-voltage electron beam which produces a thin plasma sheath on top of the melt. The low-energy secondary electrons from the plasma sheath are pulled upwards into the reaction zone by an electrode placed above the pool biased to a low positive dc or ac potential (20 to 100 V), thus creating a plasma-filled region between the electrode and the electron-beam gun. The low-energy electrons have a high ionization cross section, thus ionizing or activating the metal and gas atoms and increasing the reaction probability on collision. Charge-exchange processes between positive ions and neutral atoms take place in the plasma. In addition, as suggested by Yee (*49*), transient highly excited compound species are formed. The formation of the compound is completed most probably on the substrate from these energetic and excited transient species. The synthesis of TiC by reaction of Ti metal vapor and C_2H_2 gas atoms with a carbon-to-metal ratio approaching unity was achieved with this process (*1, 30*). Moreover, by varying the partial pressure of either reactants, the carbon-to-metal ratio of carbides could be varied (*59*) at

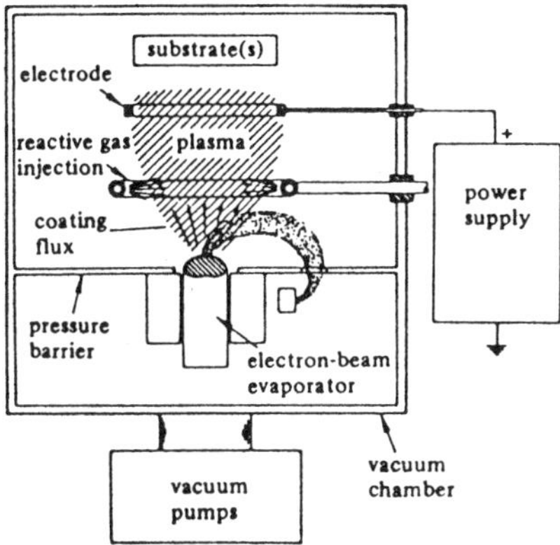

FIG. 3. The ARE process using an electron-beam evaporation source (*1*).

will. The ARE process has also been applied to the synthesis of all five different Ti–O oxides (*60*). These authors noted that in the ARE process (i.e., with a plasma) as compared to the RE process (i.e., without a plasma), a higher oxide is formed for the same partial pressure of O_2, thus demonstrating a better utilization of the gas in the presence of a plasma. The same observation was noted by Bunshah and Raghuram (*1*), as well as by Granier and Besson (*61*), for the deposition of nitrides.

b. The ARE Process Using a Resistance-Heated Evaporation Source. The basic ARE process uses electron-beam-heated sources, which are expensive and inconvenient for the evaporation of low-melting-point high-vapor-pressure materials. Nath and Bunshah (*4*) modified the ARE process for resistance-heated sources, as shown in Fig. 4. The metal vapors are generated from the chamber, the reaction being enhanced by a plasma generated by injecting low-energy electrons from a heated thoriated tungsten emitter towards a low-voltage anode assembly. A transverse magnetic field is applied to cause the electrons to go into a spiral path, thus increasing the probability of ionization.

2. Modification of the Basic ARE Process

The ARE process has substantial versatility since the substrate can be grounded, positively or negatively biased, or it can be allowed to float

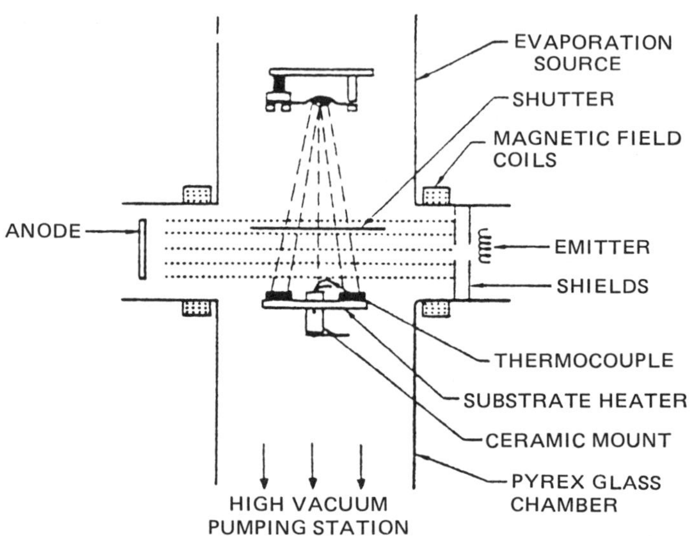

Fig. 4. The ARE process using a resistance-heated evaporation source (*4*).

electrically. There are several modifications of the basic ARE process, as illustrated in Fig. 5.

a. The Enhanced ARE Process (*3*). This is the conventional ARE process using electron-beam heating with the addition of a thermionic electron emitter (e.g., a tungsten filament) for the deposition of refractory compounds at lower deposition rates as compared to the basic ARE process. The low-energy electrons from the emitter sustain the discharge, which would otherwise be extinguished since the primary electron beam (used to melt the metal) is so weak that it does not generate an adequate

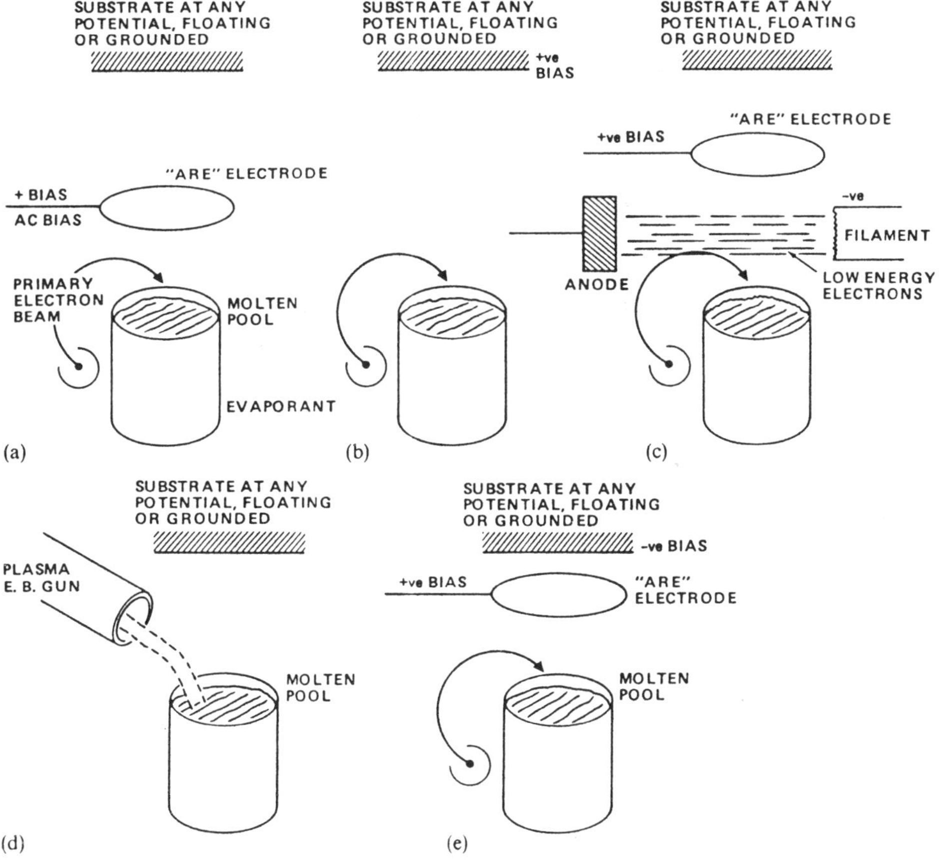

FIG. 5. (a) The basic ARE process and (b)–(e) later variations: (b) LPPD process; (c) enhanced ARE process; (d) ARE process using a hot hollow cathode electron-beam gun or a cold cathode discharge electron-beam gun; (e) BARE or RIP process.

plasma sheath above the molten pool from which the low-energy secondary electrons can be extracted by the positively biased interspace electrode. The substrate may be biased, grounded, or floating.

b. Low-Pressure Plasma Deposition (LPPD) Process. Using electron-beam evaporation sources, the electric field may be generated by biasing the substrate positively instead of using a positively biased interspace electrode. In this case, it is called low-pressure plasma deposition (LPPD) by Nakamura *et al.* (*5*). However, this version has a disadvantage over the basic ARE process since one does not have the freedom of choice to ground the substrate, let it float, or bias it negatively (the BARE process—see Subsection d below).

c. Processing Using Plasma Electron-Beam Guns. The plasma electron-beam gun, instead of the thermionic electron-beam gun, can be used to carry out the ARE process. The hot hollow cathode gun has been used by Komiya *et al.* (*62*) to deposit TiC films, whereas Zega *et al.* (*63*) used a cold cathode discharge electron-beam gun to deposit titanium nitride films. The plasma EB sources produce an abundant supply of low-energy electrons for the ARE-type process.

d. Reactive Ion Plating (RIP) Processes. If the substrate is biased in the ARE process, it is called biased activated reactive evaporation (BARE). This bias is usually negative to attract the positive ions in the plasma. The BARE process has been reinvented and called reactive ion plating by Kobayashi and Doi (*6*). Reactive ion plating (RIP) is very similar to the reactive evaporation process in that metal atoms and reactive gases react to form a compound aided by the presence of a plasma. Since the partial pressure of the gases in reactive ion plating are much higher ($>10^{-2}$ torr) than in the ARE process (10^{-4} torr), the deposits can become porous or sooty. The plasma cannot be supported by lower pressure in the simple diode ion plating process; therefore, Kobayashi and Doi (*6*) introduced an auxiliary electrode biased to a positive low voltage (as originally conceived for the ARE process) to initiate and sustain the plasma at low pressure ($\sim10^{-3}$ torr). This is no different from the ARE process with a negative bias on the substrate reported (*64*) much earlier by Bunshah, which was designated by him as the biased ARE or BARE process.

Another variation of reactive ion plating using a triode configuration (*65*) involves injection of electrons into the reaction zone between the electron-beam-heated evaporation source and the negatively biased substrate from a heated tungsten filament transversely to the metal vapor path. These low-energy electrons are pulled across the reaction zone by a positively biased anode located opposite to the cathode. The arrangement is very similar to that shown in Fig. 3 except for the use of an electron-beam-heated evaporation source and is also very similar to the triode sputtering. This adds versatility as well as complexity to the process through the addition of another process variable.

Murayama (*66*) uses an electron-beam-heated source with a negatively biased substrate and rf activation of the reactants by means of a coil electrode of aluminum wire in the reaction zone to deposit oxide and nitride films.

e. ARE Process Using an Arc Evaporation Source. Evaporation of metals using a low-voltage arc in the presence of a plasma and a negatively biased substrate is used by Snaper (*67*) and Dorodnov (*68*) to deposit nitride and carbide films, with N_2 and hydrocarbon reactive gases, respectively.

VII. Compounds Synthesized by the ARE Process and the Effect of Process Variables

The following compounds have been synthesized by the ARE process and their structure and properties studied.

Oxides: α-Al_2O_3 (*69*) and γ-Al_2O_3 (*25*), Y_2O_3 (*1, 70*), Ti oxides (*60*), TiO_2 (*24, 24a*), In_2O_3 (*4*), In_2O_3(Sn) (*4*), SnO_2 (*24a, 71*), BeO (*24a*), SiO_2 (*24a*)

Carbides: TiC (*1, 30, 72a*), ZrC (*1, 72a*), NbC (*1*), Ta_2C and TaC (*73*), VC (*1*), W_2C (*64*), HfC (*1, 72a*), VC–Tic (*74*), TiC–Ni (*75*)

Nitrides: Ti_2N and TiN (*1, 5, 61–63, 76, 77*), MoN (*68*), cubic BN (*78*), HfN (*79*), ZrN (*79*)

Sulfides: $Cu_xMo_6S_8$ (*80*), Cu_xS (*81*)

A-15 Compound: Nb_3Ge (*82*)

Carbonitrides: Ti(CN) (*83*)

Table II, from Bunshah and Raghuram (*1*), lists the lattice parameters, carbon-to-metal ratio, and the microhardness data of several carbide deposits produced by the ARE process. Also from Bunshah and Raghuram (*1*) is Table III, which shows that for a given titanium concentration in the vapor phase (i.e., constant evaporation rate) as the supply of C_2H_2 is increased, the deposit goes from Ti + TiC to TiC, whose carbon-to-metal ratio increases from 0.53 to 0.95. The corresponding variations in lattice parameter and microhardness values are also given in Table III. This illustrates the kinetic mechanism advanced earlier (Section V,1), namely that the reaction itself is no longer a rate-limiting step and therefore the

TABLE II

LATTICE PARAMETER, [C/*M*] RATIO, AND MICROHARDNESS OF CARBIDES OF GROUP IV AND V METALS (*1*)

Run no.	Carbide	Lattice parameter (Å)	Carbon/ metal ratio [C/*M*]	Microhardness (kg/mm²): Load 50 g DPHN	Microhardness (kg/mm²): Load 50 g KHN	Microhardness (kg/mm²): Published data
Ti–C_2H_2–34	TiC	4.3283	0.8—0.95	2775		2000–2750 KHN at 50 g (*49*)
Zr–C_2H_2–1	ZrC	4.7029	0.8	1285	2100	2360—unspecified load and indentor (*50*)
Hf–C_2H_2–2	HfC	4.6403	0.9—1.0	1730	2260	2276 KHN at 100 g (*51*)
V–C_2H_2–5	VC	4.1492	0.8	1924	2350	2850, DPHN, 50 g (*51*)
Nb–C_2H_2–2	NbC	4.4628	0.87	1930	2300	2400, DPHN, 50 g (*51*)
Ta–C_2H_2–2	TaC	4.4537	1.0	1285	2100	1800 DPHN, 50 g (*51*)

stoichiometry (carbon-to-metal ratio) of the deposit depends on the relative amounts of the reactants. If the quantity of C_2H_2 is increased still further, the deposit now contains TiC + C. Similar data for ZrC and (Hf–3Zr)C deposits are reported by Raghuram *et al.* (*59*).

TABLE III

VARIATION OF LATTICE PARAMETER [C/*C*] Ratio, and Microhardness with Pressure of Reactive Gas for TiC (*1*)

PC_2H_2 (torr)	Rate of evaporation (g/min)	Rate of deposition (μm/min)	Lattice parameter (Å)	Carbon to metal ratio [C/*M*]	Microhardness (50 g load) DPHN (kg/mm²)
1×10^{-4}	0.67	4	—	Ti + TiC	—
3×10^{-4}	0.67	4	4.3002	0.53	2000
4×10^{-4}	0.67	4	4.3155	0.65	2550
5×10^{-4}	0.67	4	4.3214	0.69	2775
$7\text{–}8 \times 10^{-4}$	0.67	4	4.3286	0.8–0.95	2670

In systems containing more than one stable phase, as the pressure of one of the reactants is increased holding the other constant, the deposit composition varies in accordance with the equilibrium phases shown in the equilibrium diagram. For example, in the Ti–N system, keeping the evaporation rate constant and increasing the partial pressure of N_2 gas, the deposit will consist of the following in succession:

$$\alpha\text{-Ti (with N in solid solution)}/\alpha\text{-Ti} + Ti_2N/Ti_2N + TiN/TiN$$

Furthermore, the [N/Ti] ratio in TiN will increase as the N_2 partial pressure increases.

VIII. Microstructure, Preferred Orientation, and Mechanical Properties of Refractory Compound Deposits

1. Microstructure

The microstructure and morphology of thick single-phase films have been extensively studied for a wide variety of metals, alloys, and refractory compounds. The structural model was first proposed by Movchan and Demchishin (*25*), as shown in Fig. 6, and was subsequently modified by Thornton (*84*), as shown in Fig. 7. Movchan and Demchishin's diagram was arrived at from their studies on deposits of pure metals and did not include the transition zone of Thornton's model, Zone T, which is not prominent in pure metals or single-phase alloy deposits but becomes quite

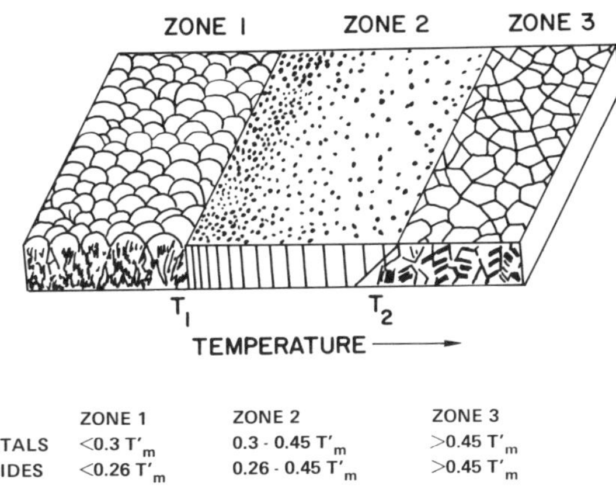

FIG. 6. Structural zones in condensates at various substrate temperatures. [After Movchan and Demchishin (*25*).]

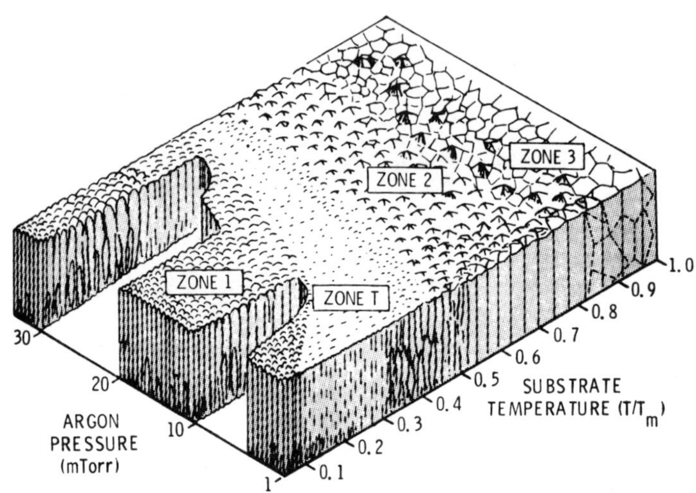

FIG. 7. Structural zones in condensates. [After Thornton (*84*).]

pronounced in deposits of refractory compounds or complex alloys produced by evaporation and in all types of deposits produced in the presence of a partial pressure of inert or reactive gas, as in sputtering or ion plating processes.

The evolution of structural morphology is as follows:

At low temperatures, the surface mobility of the adatoms is reduced and the structure grows as tapered crystallites from a limited number of nuclei. It is not a full-density structure but contains longitudinal porosity of the order of a few hundred angstroms width between the tapered crystallites. It also contains a high dislocation density and has a high level of residual stress. Such a structure has also been called "botryoidal" and corresponds to Zone 1 in Figs. 6 and 7.

As the substrate temperature increases, the surface mobility increases and the structural morphology first transforms to that of Zone T, i.e., tightly packed fibrous grains with weak grain boundaries, and then to a full-density columnar morphology corresponding to Zone 2 (Fig. 7).

The size of the columnar grains increases as the condensation temperature increases. Finally, at still higher temperatures, the structure shows an equiaxed grain morphology, Zone 3. For pure metals and single-phase alloys, T_1 is the transition temperature between Zone 1 and Zone 2 and T_2 is the transition temperature between Zone 2 and Zone 3. According to Movchan and Demchishin's original model (*25*), T_1, is 0.3 T_m for metals,

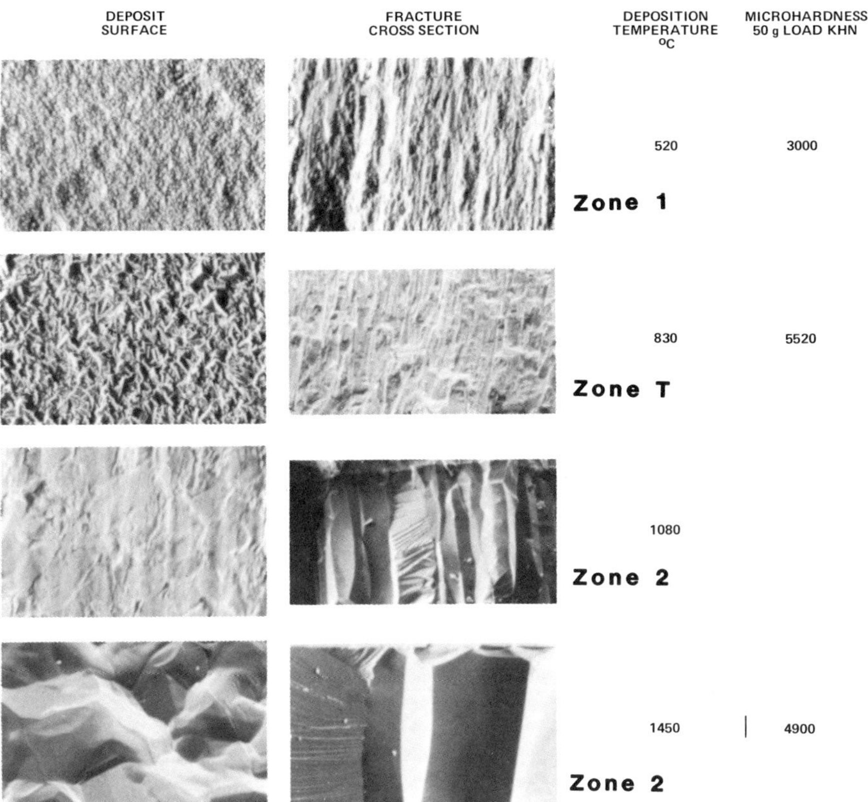

FIG. 8. Structure and microhardness of TiC deposits at various substrate temperatures (1000×). [After Bunshah *et al.* (*30*).]

and 0.22–0.26 T_m for oxides, whereas T_2 is 0.45–0.40 T_m for both (T_m is the melting point in K).

Thornton's modification shows that the transition temperatures may vary significantly from those stated above and in general shift to higher temperatures as the gas pressure in the synthesis process increases.

Optical and scanning electron microscope (SEM) studies showed that the microstructure and refractory compounds such as Ti oxides (*60*), yttrium oxide (*70*), and titanium carbide (*30*) follow the structure model of three zones (Zones 1, 2, and 3) proposed by Movchan and Demchishin (*25*). Figure 8 shows a series of microstructures illustrating the deposit surface and fracture cross section of TiC deposits produced by the ARE process with the deposition temperature varying from 520 to 1450°C. At 250°C, the microstructure shows a domed appearance characteristic of the Zone 1 structure. The fracture cross section does not resolve any details since the grain size is very fine ($<<1$ μm). At 1080 and 1450°C, the structure is clearly columnar, corresponding to Zone 2. The 830°C deposit represents the Zone T structure.

Recently, transmission electron microscopy studies carried out by Jacobson *et al.* (*85*) revealed unusual features. For TiC depositions produced at low temperature, i.e., below 700°C, the microstructure consists of an extremely fine-grained fibrous bundle structure of about 10 nm diameter with the grains separated by a fine network of cavities 1 nm in width. This is the Zone 1-type microstructure. At higher deposition temperatures, 700 to 1000°C, a more normal grain structure of about 1 μm grain diameter containing some large grain boundary cavities is produced. This is the Zone 2-type structure. On the other hand, for TiN and Ti_2N deposits (*72*) the grain size revealed by TEM is more normal and increases from 0.5 to 20 μm by increasing the deposition temperature from 550 to 1000°C.

Earlier work (*59*), based on line broadening using x-ray diffraction measurements, had reported TiC deposits produced at low deposition temperatures (500°C) had a grain size of <1000 Å with a subgrain size of 200 Å, which confirms the results reported in the TEM studies.

The effect of annealing on a low-temperature TiC deposit (*72*) produced some interesting microstructural changes. The microstructure of TiC deposited at 600°C represents that of a Zone T fibrous structure (*85*). It exhibits extremely fine columnar grains of diameter 100 Å clustered together in larger areas (700 Å), which represent a collection of grains with similar growth orientations but different rotational orientations. Such areas show up as dark contrast when their internal grains are near the Bragg diffraction orientation (Fig. 8); thus, tilting of the sample in the TEM under the electron beam would cause other areas to appear dark in the image and the previously dark areas to appear light. A typical feature of this structure is also a fine network of cavities of the order of 10 Å in width and 100 Å in mesh diameter. Figure 9 is exposed in the underfocused condition to reveal such fine cavities appearing as white lines; their contrast disappears in the focused condition. The network is irregularly distributed in this structure.

Annealing was done in a stepwise manner with 100°C intervals starting from 400°C and ending at 1000°C. The annealing time was 30 min at each step and 4 h at 1000°C. The structure was continuously observed at magnifications between 100,000 and 280,000×. No changes occurred in the image until 1000°C was reached. At that temperature the structure began to recrystallize after about 10 min, as revealed by an increasing number of

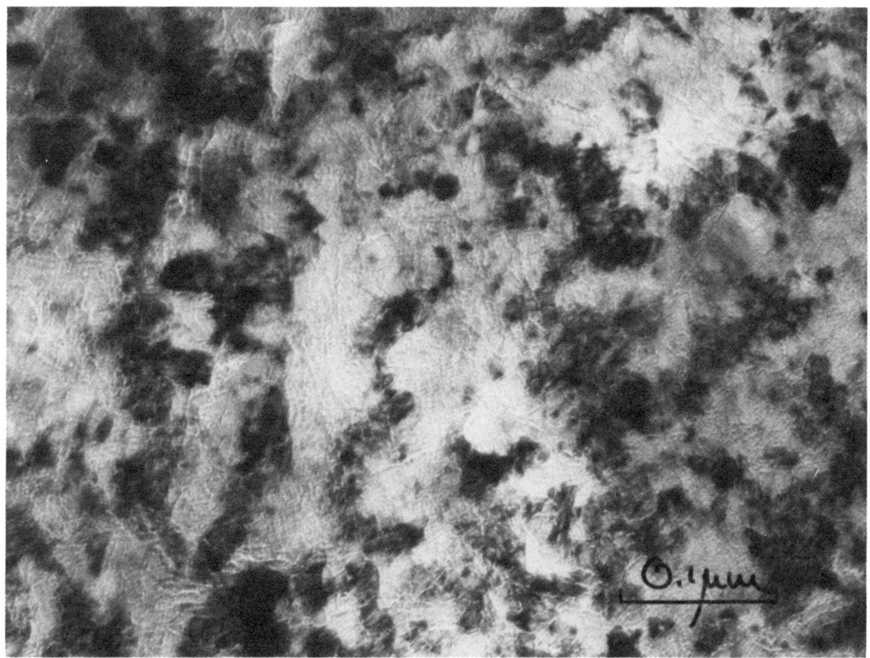

FIG. 9. TiC structure deposited at 600°C and annealed at 400°C for 30 min., showing exactly the same structure as in the as-deposited condition (*72*).

fine equiaxed grains nucleating and growing into the fibrous matrix during the following 4 h under observation. The individual grains produced a more distinct contrast (black spots) in the image. Figure 10a is taken after 3 h annealing and Fig. 10b after 4 hours; the former, in an overfocused condition, shows the cavity network in black contrast and the latter, in an underfocused condition, shows them as white lines. During recrystallization, a simultaneous coalescence of the cavities was also observed. After 4 h annealing, the mesh diameter reached approximately 200 Å, and the diameter of the equiaxed grains was 150 Å.

It is interesting to note from the above that the presence of the ultrafine cavities stabilized the microstructure, inhibiting grain growth even at the very high temperature at which recrystallization occurs.

2. Preferred Orientation

In general, deposits produced by PVD techniques tend to exhibit strong preferred orientation. The texture changes with process parameters such

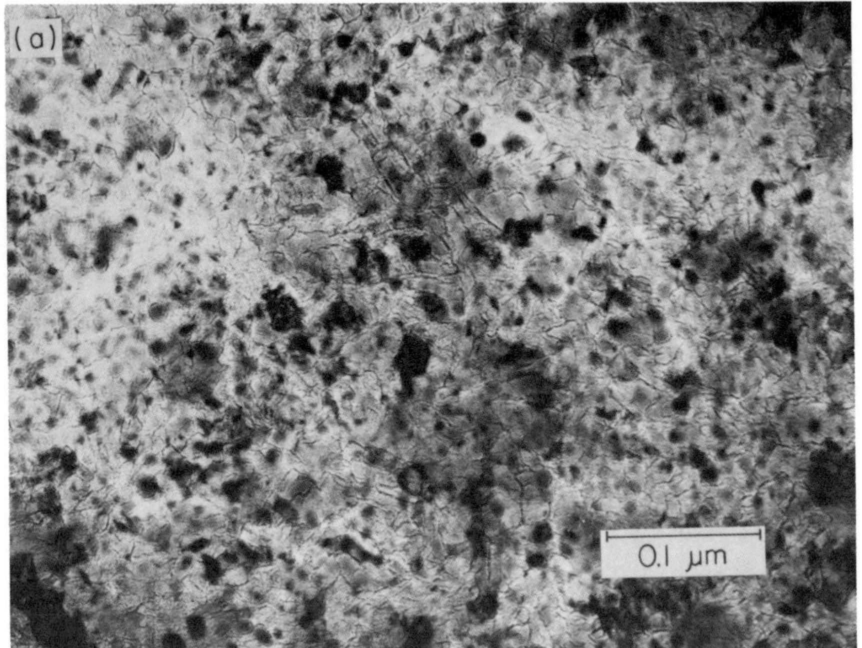
(a)
0.1 μm

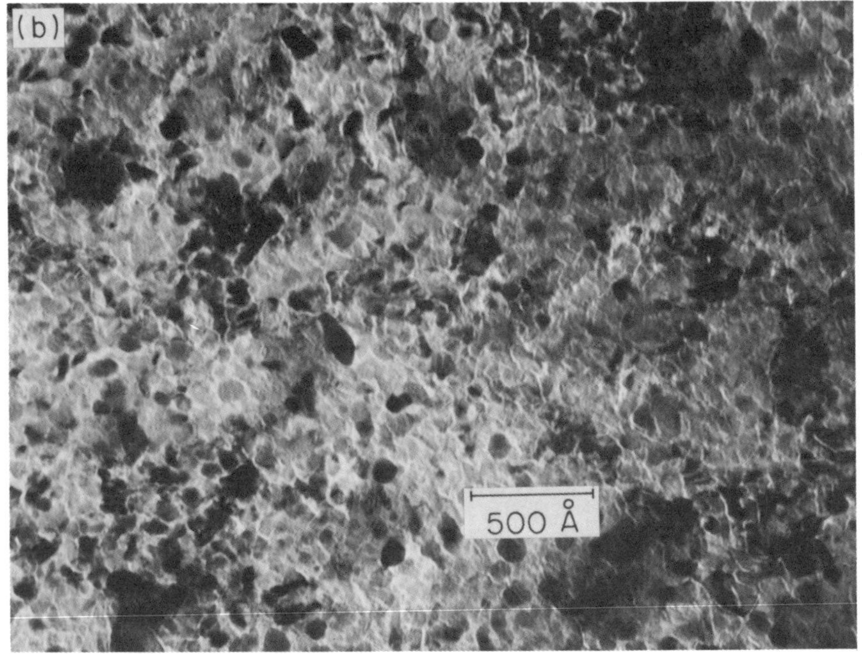
(b)
500 Å

as the deposition temperature, ion bombardment of the growing film, etc. Figure 11, taken from Raghuram and Bunshah (*30*), shows the variation in preferred orientation of TiC deposits as a function of deposition temperature. At low temperatures the deposit has a strong (220) orientation, which changes to a mixture of (200), (220), and (311) at higher temperatures.

3. Mechanical Properties

As an initial comment, one might note that most films are used in conjunction with the substrate, e.g., a hard TiN film on a cutting tool. Therefore, we are in effect dealing with the mechanical properties of a materials system consisting of a film, the interface, and the substrate and not the film *per se* as an isolated entity. This may result in film properties significantly different from bulk properties.

Refractory compound films and coatings are deposited by a variety of physical as well as chemical vapor deposition techniques. As mentioned earlier, the mechanical properties are sensitively dependent on the impurity content, stoichiometry, preferred orientation, structure, and defects. These factors can vary extensively for films synthesized by different techniques and deposition conditions. It therefore becomes difficult to generalize the results reported on these coatings. The properties of refractory carbide and nitride coatings together with available data on respective deposition conditions and structure of these films have been compiled by Bunshah *et al.* (*86, 87*) The following sections will comment on these data in order to elucidate the structure–property relationships that exist in these coatings.

The stoichiometry of the film can significantly affect the composition of the coatings and hence their properties. It is known that most of the refractory metal carbides and nitrides exist over a broad compositional range. For example, the [C/Ti] ratio for the TiC phase can vary from 0.5 to 1.0.

Let us now review the data on mechanical properties of refractory compound films.

a. Nitrides. TiN is the most widely used coating for tribological applications as well as for passivation layers in microcircuits. Other nitrides such as HfN are used to a smaller extent. The only mechanical property data available are microhardness measurements at room temperature as com-

Fig. 10. The same structure as in Fig. 9 after additional annealing at 1000°C for (a) 3 h and (b) 4 h, when the recrystallization is completed (*72*).

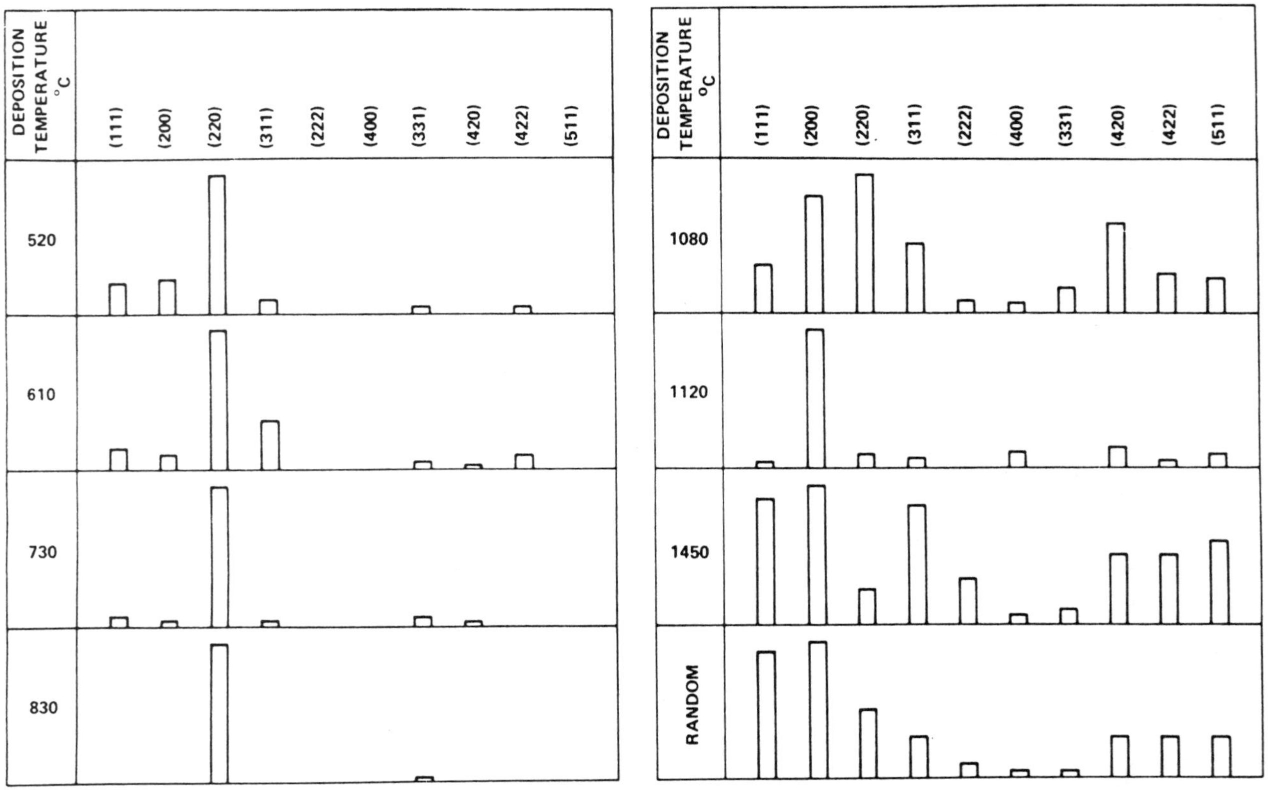

Fig. 11. Relative intensities of x-ray diffraction peaks from TiC deposits at different substrate temperatures (*30*).

TABLE IV

HARDNESS (KHN) OF TiN FILMS (*88*)

$Ti + Ti_2N$	$Ti + Ti_2N + TiN$ or $Ti + TiN$	$Ti_2N + TiN$	TiN
1160–1270		2500–2800	465, 750
		1300–2800	340–1900
1000		200–2300	1700–2200
600–700		2000–2800	1200–2000
1100–1400		1700	1700–2200
	2800–3100		400–4000
		2500	1800
			1300–3500
			2400
			2500
			<2000
			700–3100
1400–2800		1800–2500	1400–1800
		2400–3150	2100–2800
			1800–2600
			2300

piled by Sundgren (*88*) from various sources and as shown in Table IV as a function of phases present for TiN_x coatings. One can note the following:

(1) There is an enormous scatter in hardness values.
(2) There is a marked decrease in hardness value for [N/Ti] ratios <1.
(3) There is no systematic correlation with preferred orientation, deposition methodology, phases present, and microstructure.

The dependence of mechanical properties, especially microhardness, on grain size, morphology, and defect density for TiN films has been reported by many authors. In general, the coatings show the following three characteristic morphologies depending on the growth conditions: (1) coarse grained with high-angle grain boundaries, (2) coarse grained with high density of defect structure consisting of subgrains and dislocation cell boundaries, and (3) extremely fine grained. The fine-grained morphology exhibits the highest hardness, whereas the coarse-grained structure with voids and grain boundaries exhibits lower hardness. Jacobson *et al.* (*76*) have observed hardness values as low as 625 KHN for TiN coatings with such morphologies.

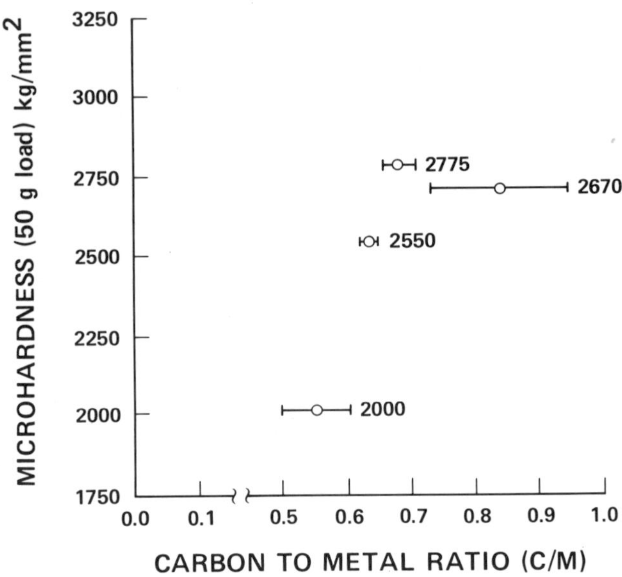

FIG. 12. Microhardness versus carbon-to-metal ratio for TiC coatings prepared using the activated reactive evaporation process (*1*).

b. Carbides. Figure 12 shows the variation in microhardness of TiC_{1-x} coatings prepared by the ARE technique as reported by Bunshah and Raghuram (*1*). It is seen that the hardness increases with [C/Ti] ratio. Also, there is reasonably good agreement between the properties of the bulk and film, considering the fact that the preferred orientation of the various specimens is not specified. The influence of preferred orientation on the hardness of TiC films produced by ARE (*30*) is given in Fig. 13. The variation in preferred orientation is given in Fig. 11. It can be seen from this figure that the coatings deposited below 1100°C show a strong (220) preferred orientation, whereas the preferred orientation of the coatings deposited above 1100°C is (200). Correspondingly, the hardness also shows a significant increase from 2600 KHN for (220)-oriented coatings to about 500 KHN for (200)-oriented coatings.

This progressive increase in hardness with changes in microstructure is also observed in oxide coatings, as will be illustrated later. At the low deposition temperature, the deposit is very fine grained with a network of cavities so that the material can deform more easily than the one deposited at higher temperatures where the grain size is larger but the fine cavity network has disappeared. The hardness values at the high deposi-

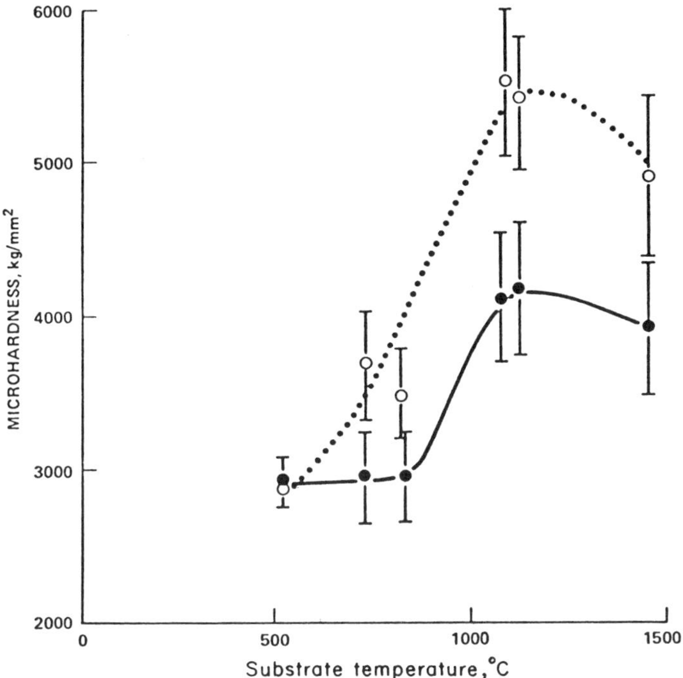

FIG. 13. Variation of microhardness with deposition temperature for TiC *(30)*. (●), DPHN; (···), KHN. 50 g load.

tion temperatures are very high and approach that of diamond. It should be pointed out that hardness values will vary considerably with the direction in which they are measured, i.e., on the surface or in the cross section, since the deposits have a strong preferred orientation. It is not possible to separate the relative contributions of the microstructure and preferred orientation to microhardness.

c. Carbonitrides. A systematic study relating the composition, microstructure, preferred orientation, deposition temperature, and microhardness for the titanium carbonitrides from pure TiC and Ti(CN) to TiN was recently reported by Jacobson *et al.* *(83)* Their results are shown in Figs. 14 and 15. The principal conclusions from the study are:

(1) The hardness increases as the grain size decreases, and features such as high-angle grain boundaries, pores, and cracks are eliminated.

(2) The hardness increases with a change in preferred orientation from (220) to (200).

HARDNESS, KHN_{50g}

	500°C	350°C	250°C
TiN	2,460	3,040	3,040
70:30	3,630	4,700	4,020
50:50	4,410	5,200	4,480
30:70	4,550	5,880	4,940
TiC	5,880	5,570	4,700

FIG. 14. Hardness variation in Ti(N,C) coatings deposited on high-speed tool steels at various composition and deposition temperatures (*83*).

(3) Again, it is not possible to separate out the contributions of each of these factors.

d. Oxides. Shown in Fig. 16 is the variation in microhardness with deposition temperature for Al_2O_3 and ZrO_2 observed by Movchan and Demchishin (*25a*). The increase in hardness when the deposition temperature increases from $0.3T_m$ observed for ZrO_2 and Al_2O_3, unlike metals, is attributed by the authors to a "more perfect" material produced at high deposition temperature due to the "volume processes of sintering." A similar hardness variation was obtained by Colen and Bunshah (*70*) for Y_2O_3 film, as shown in Fig. 17. It should be noted that the absolute values

GRAIN MORPHOLOGY

		500°C	350°C	250°C
TiN	(220)	3,000 Å HGb Cracks	2,000 Å HGb	-
70:30	(220)	1,000 Å HGb No pores	3,000 Å Subgrains	-
50:50	(220)	3,000 Å Subgrains and cells	100 Å Pores	-
30:70	(220)	100 Å No pores	-	-
TiC	(200)	100 Å No pores	50 Å Cells	-

TEXTURE – TiC PRESENT WORK TiC(200) – KHN 5,880
PRIOR WORK TiC(220) – KHN 3,000

FIG. 15. Grain morphology, defect distribution, and texture in Ti(N,C) coatings deposited on high-speed tool steels at various composition and deposition temperatures (*83*).

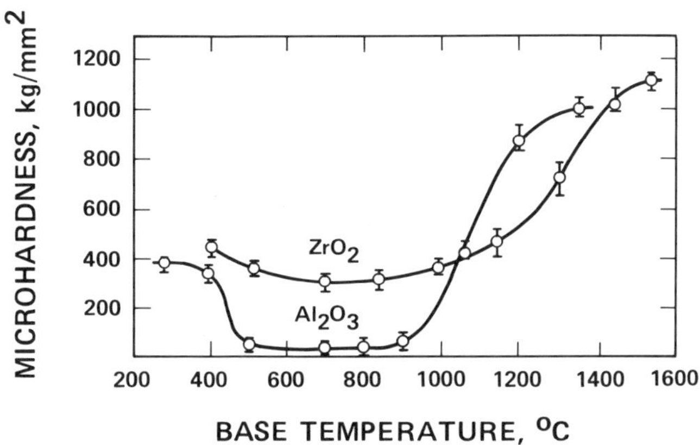

FIG. 16. Microhardness of condensate versus base (substrate) temperature (*25a*).

of microhardness for the oxides are much lower than those produced by sintering at high temperature or by CVD, which also involves high deposition temperatures.

Colen and Bunshah (*70*) have also demonstrated the dependence of the fracture behavior Y_2O_3 coatings on the grain size. They used the Hertzian fracture test to measure the fracture stress and surface energy at the

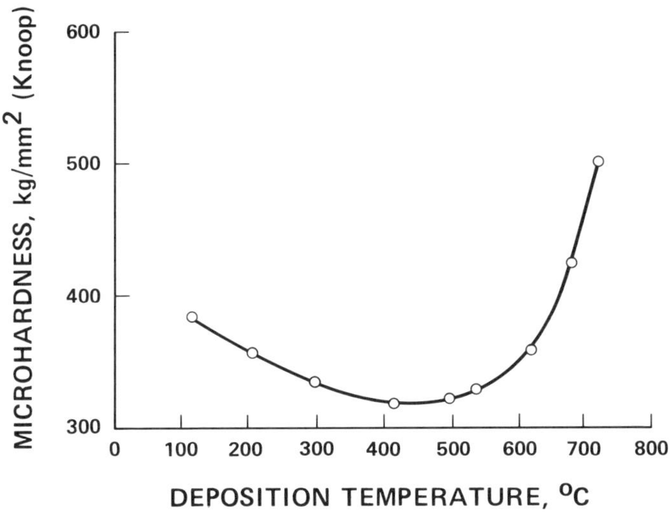

FIG. 17. Microhardness versus deposition temperature for Y_2O_3 films (*70*).

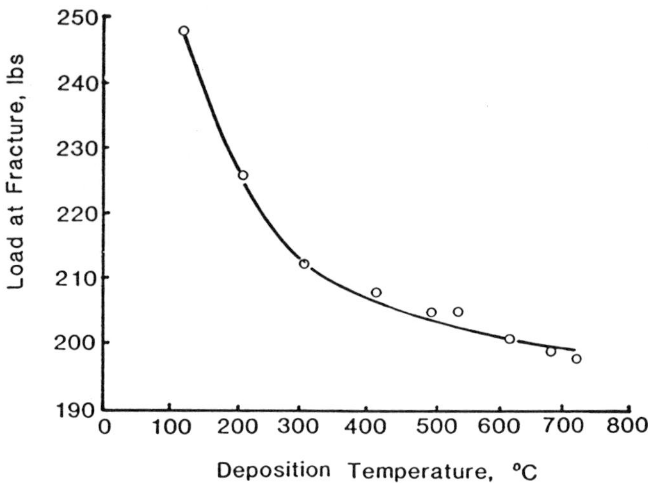

FIG. 18. Load at fracture versus deposition temperature for Y_2O_3 films (*70*).

fracture surface. As shown in Fig. 18, they observed that the fracture stress decreases with an increase in deposition temperature. This decrease can be attributed to the increase in grain size, which consequently decreases the surface energy.

IX. Applications of the ARE Process

In general, coatings are used in decorative, optically functional, mechanically functional, chemically functional, and electrically functional applications. Examples of applications falling into these categories are given below.

1. TELECOMMUNICATIONS EQUIPMENT

Yoshihara and Mori (*3*) used the enhanced ARE process to deposit TiN films onto telecommunication equipment components.

2. LOCALIZED CORROSION RESISTANCE

Agarwal *et al.* (*89*) showed that TiC and TiN coatings on type M-50 high-speed steel eliminated the sever-localized corrosion observed on this material in an engine oil environment containing 3 ppm chlorine. This is particularly important for bearing applications, since the hard coatings will confer both corrosion and wear resistance.

3. Superconducting Compounds

Superconducting Chevral-phase copper molybdenum sulfide Cu_xMo_6 films were deposited by reactive evaporation and activated reactive evaporation using H_2S as the reactive gas with coevaporation of Mo and Cu from two sources (*80*).

In subsequent work, the *A*-15 compound Nb_3Ce was deposited by the ARE process by electron-beam evaporation of Nb in the presence of a low partial pressure ($\sim 10^{-4}$ torr) of germane gas (*82*). This is the first report to the best of our knowledge for the deposition of *A*-15 compounds using reactive evaporation-type processes. At low deposition temperatures (~450°C) the deposit is amorphous or microcrystalline and can be readily bent. It crystallizes to the Nb_3Ge phase on annealing at 850°C for 1 h in vacuum.

4. Transparent Conducting Coatings

Such coatings require a combination of low electrical resistivity and wide-band optical transmission from 0.4 to 1.6 μm for solar cell optoelectronic and display applications. Nath and Bunshah (*4*) deposited indium oxide and tin-doped indium oxide films using the ARE process. The properties are dependent on deposition temperature, which was varied from room temperature to 350°C for lowest-resistivity films. This process has a major advantage over others such as spray pyrolysis and reactive sputtering since no post-deposition annealing (reactive sputtering) or very high substrate temperatures (spray pyrolysis) are required. Deposition rates can be as high as 0.04 μm/min. Films with electrical resistivity of 7×10^{-5} Ω cm with an integrated (0.4–1.2 μm) transmittance of over 90% were obtained (*90*). Thicker films have a sheet resistance as low as 2.2 $\Omega/\square$ (*4*).

5. Amorphous Silicon for Photovoltaic Applications

Amorphous hydrogenated silicon (*a*-Si : H) alloy semiconductors, prepared by glow discharge dissociation (GD) of silane and also by reactive sputtering (RS) of Si in hydrogen plasma, have become promising materials for the fabrication of thin-film photovoltaic cells. Apart from photovoltaic applications, *a*-Si : H is a promising material for other commercial applications such as xerography, logic circuits, imaging devices, etc. However, the economic feasibility of *a*-Si : H devices is strongly dependent on the high throughputs which are necessary to fabricate large-area devices at considerably lower prices (*91*). The limitations inherent due to interdependence of the process parameters which control the deposition rate and the plasma conditions in case of glow discharge and reactive sputtering processes make is difficult to obtain deposition rates higher than a few angstroms per second (typically 1–2 Å/sec) by these techniques. Hence neither of the above processes is suitable for production of large-area devices. It is therefore necessary to look for alternative techniques which would be capable of providing high deposition rates to meet high-throughput production requirements. Activated reactive evaporation offers such an alternative.

The process of synthesis of *a*-Si : H by ARE involves electron-beam evaporation of high-purity silicon in the presence of a radio frequency excited by hydrogen plasma. Unlike the glow discharge and reactive sputtering processes, ARE has the unique capability for independent control over the evaporation rate, plasma intensity, and location of the substrates with respect to the plasma, which are important for controlling the properties of the deposited *a*-Si : H material. These features enable us to control the material. The important variables during the deposition process are the evaporation rate, hydrogen partial pressure, plasma intensity, substrate temperature, and substrate bias. Doped films can be prepared by evaporating Si in the plasma of H_2 plus the dopant gas. PH_2 and B_2H_6 gases will be used for *n*- and *p*-type doping, respectively.

The applicability of the ARE technique for synthesizing amorphous hydrogenated silicon films has already been demonstrated by Anderson an co-workers (*92*). Work is also being carried out in our laboratory on high-rate deposition of *a*-Si : H films by the ARE process. The films deposited at ≃150 Å/sec with dc excitation of the plasma at 2 mtorr hydrogen pressure show conductivities in the range of $10^{-10}\ \Omega^{-1}\ cm^{-1}$ to $10^{-11}\ \Omega^{-1}\ cm^{-1}$ with an order of magnitude change in the photoconductivity. The hydrogen concentration in such films as determined by elastic scattering of the hydrogen by α particles and also by infrared measurements is close to 15%. The hydrogen–silicon bonding modes are mono- and dihydride type. The silicon dangling bonds associated spin-density measurements carried out on these films at the Xerox Palo Alto Research Center showed 2×10^{18} spins/cm^3. Another set of experiments directed towards further improvements in these properties involved a variation of deposition parameters such as the rate (15–500 Å/sec), rf excitation of the plasma, and rf/dc substrate bias. The efforts have led to a photoconductive gain of three orders of magnitude at a rate ~25 Å/sec. The results are summarized in Table V (*93*). Although the photoconducting ratios in our films deposited using the ARE technique are lower than those observed in

TABLE V

PRELIMINARY DATA ON *a*-Si : H DEPOSITED BY ARE AT UCLA (*93*)

a. dc Excitation

C_H (%)	Lump	Deposition rate (μm/min)	dc (+)	Bias (V)	Dark conductivity (ω^{-1} cm^{-1})	Activation energy (eV)	Photoconductive ratio
5.8	Intrinsic	0.32	80	−200	4.1×10^{-8}	0.68	8
6.0	Intrinsic	0.27	80	−200	1.13×10^{-7}	0.63	8
4.0	Intrinsic	0.37	350	0	6.4×10^{-8}	1.02	4

b. rf Excitation

C_H (%)	Lump	Deposition rate (μm/min)	dc (+) (rf)	Bias excitation (ω^{-1} cm^{-1})	Dark conductivity (eV)	Conductivity energy ratio	Conductive energy (eV)
24.7	Intrinsic	0.16	250	5×10^{-10}	0.71	60	1.7
16.9	Intrinsic	0.36	250	1×10^{-9}	0.75	20	1.57
11.5	Intrinsic	0.77	250	3×10^{-9}	0.58	4	1.47
5.1	Intrinsic	1.67	250	5×10^{-9}	0.50	1	1.37

a-Si : H films prepared by glow discharge, the properties of our films are better than the best quality sputtered *a*-Si : H alloys deposited at a rate of 2–3 Å/sec. The lower values of photoconductivity in our films as compared to GD films are believed to be due to a high level of impurity content in the films. The vacuum system used in these studies had a high background pressure, which is believed to be the source of the impurities. The existing system is being modified to ensure a low level of residual impurities. We believe higher photoconductivity would be obtained by reducing the impurity concentration in our films.

6. Reversible-Cathode Materials for Solid Electrolyte Batteries

A variety of Group IVB, VB, and VIB transition-metal dichalcogenides have attracted considerable interest as reversible cathodes in ambient-temperature secondary batteries because of their high diffusivity for alkali ions (*94*). Among the chalcogenides investigated as secondary lithium battery cathodes, the layered chalcogenide TiS_2 displays both reversibility and a high energy density (*95*). Its high electrical conductivity, lithium diffusivity, and electrochemical potential versus lithium are all quite favorable for battery operation. Furthermore, lithium intercalation in TiS_2 has proven to be reversible over many charge–discharge cycles in laboratory experiments.

The ARE technique represents a suitable method for synthesis of transition-metal dichalcogenide films, for it offers excellent control over the stoichiometry of the films, which in turn affects the performance of the battery. TiS_2 films have been prepared in our laboratory using the ARE technique where Ti was evaporated using an electron beam in H_2S plasma. The critical deposition parameters which controlled sulfur content in the film and the resultant sulfide phases were H_2S flow rate and substrate temperature (*96*). The films deposited at 400°C with a hydrogen flow rate of 400 SCCM exhibited the bias–electrochemical properties. The [Ti/S] ratio in these films was comparable to the standard TiS_2 powder, indicating that the films were stoichiometric. The films showed broad diffraction peaks, suggesting that a significant microcrystalline or amorphous content was present (a-TiS_2). The resistivity of the films was excellent, ranging from 7×10^{-4} to 2×10^{-3} Ω cm. These values compare quite well with reported literature values for TiS_2. Moreover, the films showed metallic behavior (i.e., an increase in resistivity with temperature) like bulk Ti_2.

The a-TiS_2 films produced by the ARE technique were able to both intercalate and deintercalate lithium (*97*). The open-circuit voltage of a-TiS_2 films versus lithium was 2.7 V. Films that were treated for three days with an excess of *n*-butyl lithium exhibited an open-circuit voltage of 1.7 V versus lithium. These lithium composition limits were confirmed in electrochemical experiments. Full battery discharge resulted in a potential of 1.7 V versus lithium, while complete charging produced a delithiated cathode with a potential of 2.7 V versus the Li anode. *In situ* four-wire resistivity measurements revealed very little change in the electrical resistivity of thin-film a-TiS_2 as a function of lithium content (y) in Li_yTiS_2. The electrical measurements gave no indication of any sizeable ionic contributions to the conductivity. Although no long-term cycling studies were performed, the thin-film a-TiS_2 material was charged and discharged several times with no apparent lack of capacity. Cathodes appeared unchanged after electrochemical cycling, and no cracking or exfoliation was observed.

7. Cubic Boron Nitride

Thin films of boron nitride have attracted attention because of their applications as wear-resistant, corrosion-resistant, insulating, and passivating films that can be used in high-temperature and/or corrosive environments. Several chemical (*98–107*) as well as physical vapor deposition processes (*108–110*) have been used to deposit these films. However, most of these techniques are expensive and involve the use of toxic gases such as hexachloroborazate, trichloroborozate or borozine. A simple inexpensive process for the synthesis of CBN was developed by Bunshah and Deshpandey in collaboration with Chopra *et al.* (*111*). In this process, called activated dissociation reduction reaction (ADRRP), boric acid is evaporated using a resistance-heated boat in an NH_3 plasma (*78*). Some of the results on CBN films prepared using the above technique are given below.

The rate of deposition of the boron nitride films was measured to be between 100 and 150 nm min^{-1}. The identification of the deposited coating as boron nitride was primarily based on the observed ir spectra and electron diffraction patterns. Figure 19 shows the ir spectrum of a coating on silicon deposited at 450°C. The strong absorption band at 6.8 μm corresponds to the B–N bending vibration (*101–103, 105, 112*), and another sharp peak near 11.5 μm is due to the B–N–B bending vibration (*101,*

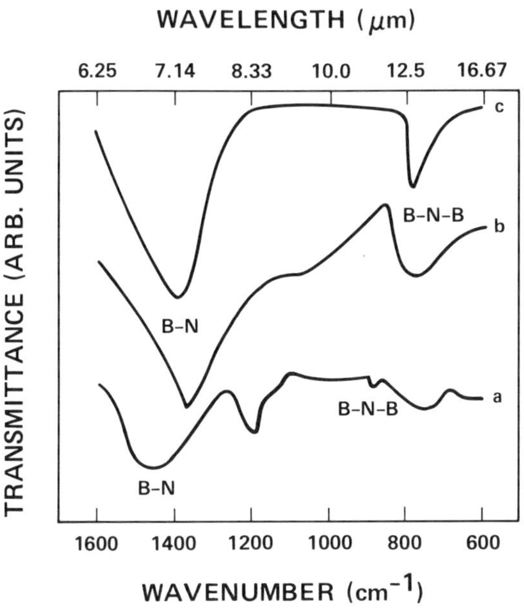

FIG. 19. IR spectra of cubic boron nitride coatings: Curve a was obtained at 450°C by Chopra *et al.* (*111*); curve b, at 1150°C by Takahashi *et al.* (*101*); curve c, at 900°C by Rand and Roberts (*102*). [Taken from Chopra *et al.* (*78*).]

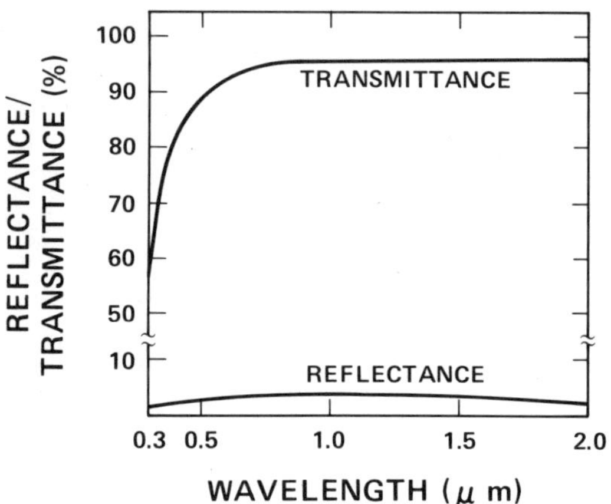

FIG. 20. Transmittance and reflectance spectra of a cubic boron nitride film of thickness 400 nm deposited at 450°C onto glass substrate (*78*).

102). The position of the absorption band for B–N vibration was found to be independent of the deposition temperature, but the magnitude of the absorption changes slightly with the deposition temperature and crystallinity, probably as a result of the change in the number of B–N bending vibrations contributing to the spectrum.

Figure 20 shows the reflectance and transmittance spectra for a film of thickness 400 nm deposited at 450°C onto a glass substrate. The band gap of the coating material was calculated from the absorption coefficient and was found to be 3.64 eV. It should be noted that Rand and Roberts (*102*) have reported a value of 3.8 eV for CVD hexagonal BN films on fused silica.

Films deposited at temperatures below 200°C were amorphous but polycrystalline films were obtained for high deposition temperature (450°C). The films deposited at 450°C showed continuous ring type electro diffraction pattern indicating polycrystalline films. The measured values for these films are shown in Table VI. These values are compared with cubic boron nitride *d* values from the ASTM data card (*113*) and also with those for a simple cubic phase reported by Voskoboynikov *et al.* (*112*). The formation of cubic boron nitride is clearly established. The calculated lattice parameter is 0.367 nm, which matches with the standard value of 0.362 nm. The presence of faint lines corresponding to the simple cubic phase indicates that this phase is present in trace amounts. A typical micrograph of this film is shown in Fig. 4b. The microstructure shows fine grains of size about 25 nm. Compositional analysis of the films by Auger

TABLE VI

LATTICE SPACING *d* OF BORON NITRIDE COATINGS COMPARED WITH STANDARD *d* VALUES (*78*)

Measured *d* (nm)	Observed intensity	Standard values for simple cubic boron nitride (112) *d* (nm)	(*hkl*)	Standard values for zincblende structure of boron nitride (113) *d* (nm)	(*hkl*)
—	—	0.362	(100)	—	—
—	—	0.257	(110)	—	—
0.210	Very strong	0.209	(111)	0.209	(111)
0.180	Medium	0.181	(200)	0.181	(200)
0.169	Very weak	0.162	(210)	—	—
—	—	0.148	(211)	—	—
0.127	Strong	0.128	(220)	0.128	(220)
0.108	Strong	—	—	0.1086	(222)
0.088	Weak	—	—	0.0904	(400)

electron spectroscopy has shown the existence of carbon and oxygen impurities with a concentration of less than 10 at. % each.

The adherence of the coatings was found to be good at deposition temperatures above 400°C. The microhardness of the coatings deposited onto stainless steel was measured with a Vickers microhardness tester at 10 g f load. The average value of the microhardness was found to be 2128 kg f mm^{-2}. This value compares well with others reported in the literature (Table VII).

8. OPTICAL COATINGS

Very recently, Ebert (*24, 24a*) has given an extensive account of the use of the ARE process for the preparation of TiO_2, BeO, In_2O_3, SnO_2,

TABLE VII

COMPARISON OF PRESENT RESULTS WITH REPORTED RESULTS (*78*)

Physical property	Present results	Reported results
Lattice parameter (nm)	0.367	0.362
Band gap (eV)	3.67	3.8
Microhardness (kg mm^{-2})	2000–2200 at 10 gf load on stainless steel	2200–3100 at 25 g load (boron nitride on TiN-coated WC) 1600–220 at 25 gf load (boron nitride on Al_2O_3-coated WC)

and SiO_2 coatings for use as high-quality optical coatings for laser mirrors and antireflection coatings.

9. Wear of Metal Surfaces

TiC and TiN films produced by the ARE process showed excellent wear resistance under adhesive, abrasive, and erosive wear conditions. In contrast to ion implantation, the wear properties are not dependent on organic lubricants; i.e., these films show low wear when run dry (*114*). Ion-implanted surfaces will show high wear when run dry. Some of the refractory compounds show a fairly low coefficient of friction (as low as 0.2 to 0.3). They are a promising approach when wear resistance is needed at temperatures greater than those where MoS_2 is stable, approximately 300°C.

Later, Jamal *et al.* (*115*) studied the friction and adhesive wear couples, where both components of the couple are refractory hard compounds such as TiC and TiN. The test apparatus was a conventional pin-on-disc tribotester.

The coefficient of friction and the wear were very much lower when the test couple consisted of a hard coating rubbing against a hard coating than when one or both of the components were uncoated metals, even in dry (unlubricated) conditions. In some cases for the hard/hard couple, a very low coefficient of friction (0.5 to 0.1) and very low wear (i.e., wear groove depth less than 1×10^{-5} in. was observed for test conditions of 500 m run distance under 0.4 kg load in dry conditions.

In the case of abrasive wear, the presence of hard overlay coatings greatly diminishes the cutting action of the abrasive particles that may be present (*114*) by a factor of 4 to 8.

In another test, the abrasive wear of TiC-coated stainless steel is 34 times better than that of hard electrodeposited chromium using Al_2O_3 wear particles (*64*).

10. Wear of Cutting Tools

The current status is outlined in the following sections.

a. Carbide cutting tools. TiC, TiN, Al_2O_3, and HfN coatings produce improvement by a factor of 2 to 8 over uncoated tools in continuous cutting only, e.g., on coated inserts; CVD and PVD processes produce equal results. In interrupted cutting, practically no improvement is evident except in two cases: (1) sputter TiN; and (2) TiC or HfN by the ARE process (*116*). No data are reported on shaped tools.

An important and significant difference between ARE and CVD coatings on cemented carbide tools was reported recently by Kodama and Bunshah (*117*) for interrupted cutting tests. The ARE coatings of HfN and TiC stood up very well, whereas the CVD coatings of the same material failed promptly. The reason is that the CVD coatings have a large-grained microstructure with no toughness, whereas the ARE coatings have a very fine-grained microstructure with the fine cavity network discussed above, which gives the coating considerable toughness and the ability to absorb the intermittent loads imposed by the interrupted cutting conditions.

b. High-Speed Steel Tools. In continuous cutting, TiC, TiN, and HfN coatings produce an improvement by a factor of 2 to 8 depending on the cutting conditions (*118, 119*). This is illustrated by Bunshah (*64*).

For drills, TiN and TiC coatings made by the ARE process have produced an improvement by a factor of 20 over uncoated drills and by a factor of 4 over the standard "black oxide" coated drills (*120*). This represents a major improvement. Such coatings are now extensively used on high-speed steel tools such as drills, hobs, gear cutters, broaches, etc., all over the world (*121*).

X. Summary

This review describes one of the original plasma-assisted reactive deposition processes (the other being the reactive sputtering process) for the production of films and coatings of refractory compounds such as oxides, carbides, nitrides, etc. In this process, metal or alloy vapors produced from evaporation sources (resistance, induction, or electron-beam heated) are brought into contact with molecules of a reactive gas in the presence of a plasma produced in the space between the evaporation source and the substrate. The reactive evaporation process is "activated" in the presence of a plasma by the production of ions and energetic neutrals, resulting in the formation of metastable species which react to form the refractory compound deposit. The process is very versatile: (1) A very large number of compounds can be produced; (2) the deposition rate can be varied over a wide range from 50 to 250,000 Å/min thickness at a normal source-to-substrate distance of 20 cm; (3) the substrate temperature can be varied over a wide range from room temperature to very high temperatures (1500°C) independently of the evaporation rate; and (4) the substrate can be protected from damage by bombardment with electrons or ions; alternately, the substrate can be biased to intentionally produce ion bombardment of the growing film to cause desired changes in microstructure, residual stresses, and properties of the film.

The process and its derivatives are used on an industrial scale in a diverse variety of applications ranging from decorative coatings to coatings for protection against wear and corrosion, and for microelectronic applications. It can also be used to produce unique microstructures such as microlaminate composites, i.e., bulk coatings with a total thickness up to 500 μm consisting of a large number of alternate layers, each of which has a thickness in the range of 1 to 5 μm. The microlaminate system can be metal–ceramic or ceramic–ceramic.

References

1. R. F. Bunshah and A. C. Raghuram, *J. Vac. Sci. Technol.* **9,** 1385 (1972).
2. R. F. Bunshah, U.S. Patent 3,791,852 (1974).
3. H. Yoshihara and H. Mori, *J. Vac. Sci. Technol.* **16,** 1007 (1979).
4. P. Nath and R. F. Bunshah, *Thin Solid Films* **69,** 63 (1980).
5. K. Nakamura, K. Inagawa, K. Tsuroka, and S. Komiya, *Thin Solid Films* **40,** 155 (1977).
6. M. Kobayashi and Y. Doi, *Thin Solid Films* **54,** 57 (1978).
7. Soddy, *Proc. R. Soc. London* **78,** 429 (1907).
8. I. Langmuir, *J. Am. Chem Soc.* **35,** 931 (1913).
9. S. Brinsmaid, W. J. Keenan, G. E. Koch, and W. F. Parsons, U.S. Patent 2,784,115 (1957).
10. M. Auwarter, U.S. Patent 2,920,002 (1960).
11. C. S. Herrick and A. D. Tevebaugh, *J. Electrochem. Soc.* **110,** 199 (1963).
12. M. A. Novice, J. A. Bennett, and K. B. Cross, *J. Vac. Sci Technol.* **1,** 73A (abstr.) (1964).
13. R. B. Schilling, *Proc. IEEE* **52,** 1350, (1964).
14. E. Ritter, *J. Vac. Sci. Technol.* **3,** 225 (1966).
15. E. Ferrieu and B. Pruniaux, *J. Electochem. Soc.* **116,** 1008 (1969).
16. J. R. Rairden, *Electrochem. Technol.* **6,** 269 (1968).
17. J. R. Rairden, *in* "Thin Film Dielectrics" (F. Vratny, ed.), p. 279. Electrochem. Soc., New York, 1969.
18. J. DeKlerk and E. F. Kelley, *Rev. Sci. Instrum.* **36,** 506 (1965).
19. A. J. Learn and K. E. Haq, *Appl. Phys. Lett.* **17,** 26 (1970).
20. M. T. Wank and D. K. Winslow, *Appl. Phys. Lett.* **13,** 286 (1968).
21. B. B. Kosicki and D. Khang, *J. Vac. Sci. Technol.* **6,** 592 (1969).
22. W. Heitmann, *App Opt.* **10,** 2414 (1971).
23. W. Heitmann, *App Opt.* **10,** 2685 (1971).
24. H. Küster and J. Ebert, *Thin Solid Films* **70,** 43 (1980).
24a. J. Ebert, *SPIE* **325,** 29 (1982).
25. B. A. Movchan and A. V. Demchishin, *Phys. Met. Metallogr.* (*Engl. Transl.*) **28,** 83, 1969.
25a. B. A. Movchan and A. V. Demchishin, *Fizika Metall.* **28,** 653 (1969).
26. D. Hoffman and D. Liebowitz, *J. Vac. Sci. Technol.* **9,** 326 (1972).

27. R. F. Bunshah, R. J. Schramm, R. Nimmagadda, B. A. Movchan, and V. P. Borodin, *Thin Solid Films* **40,** 169 (1977).
28. T. Abe, K. Inagawa, R. Obasa, and Y. Murakami, *Comm. Eur. Communities* [*Rep.*] *EUR* (1982).
29. R. F. Bunshah, ed., "Deposition Technologies for Films and Coatings." Noyes, Park Ridge, New Jersey, 1982.
30. R. F. Bunshah and A. C. Raghuram, *J. Vac. Sci. Technol.* **9,** 1389, (1972).
31. J. A. Thornton, *in* "Deposition Technologies for Films and Coatings" (R. F. Bunshah, ed.). Noyes, Park Ridge, New Jersey, 1982.
32. S. Schiller, U. Heisig, K. Steinfelder, and J. Strumpfer, *Thin Solid Films* **64,** 455 (1979).
33. S. Maniv, C. Miner, and W. D. Westwood, *J. Vac. Sci. Technol.* **18,** 195 (1981).
34. S. Maniv, C. Miner, and W. D. Westwood, *J. Vac. Sci. Techol., A***3,** 1370 (1983).
35. W. D. Sproul, U.S. Patents 4,428,811 and 4,428,812 (1984).
36. W. D. Sproul, *Thin Solid Films* **118,** 279 (1984).
37. M. Scherer and P. Wirz, *Thin Solid Films,* **119,** 203 (1984).
38. C. Deshpandey and R. F. Bunshah, *J. Vac. Sci. Technol.,* **3,** 553 (1985).
39. A. T. Bell, *in* "Techniques and Applications of Plasma Chemistry" (J. R. Hollahan and A. T. Bell, eds.), p. 31. Wiley, New York, 1974.
40. F. J. Kampas, *in* "Semiconductors and Semimetals," (J. I. Pankove, ed.) Vol. 21, Part A, p. 159. Academic Press, New York, 1984.
41. J. A. Thornton, *Thin Solid Films* **107,** 3 (1983).
42. M. J. Kushner, *J. Appl. Phys.* **53,** 2923 (1982).
43. P. A. Longeway, *in* "Semiconductors and Semimetals" (J. I. Pankove, ed.), Vol. 21, Part A, p. 195. Academic Press, New York, 1984.
44. D. W. Hess, *in* "VLSI Electronics: Microstructure Science" (N. G. Einspruch and D. M. Brown, eds.), Vol. 8, p. 53. Academic Press, New York, 1984.
45. H. A. Weikliem "Semiconductors and Semimetals" (J. I. Pankove, ed.), Vol. 21, Part A, p. 195. Academic Press, New York, 1984.
46. D. L. Flamm, V. M. Donnelly, and D. E. Ibbotson, *in* "VLSI Electronics: Microstructure Science" (N. G. Einspruch and D. M. Brown, eds.), Vol. 8, p. 189. Academic Press, New York, 1984.
47. H. Yasuda, *in* "Thin Film Processes" (J. L. Vossen and W. Kern, eds.), p. 361. Academic Press, New York, 1978.
48. C. R. Aita and T. A. Myers, *J. Vac. Sci. Technol. A* **1,** 348 (1983).
49. K. K. Yee, *Proc. Conf. Chem. Vap. Deposition, Int. Conf., 5th, 1975,* p. 238 (1975).
50. R. F. Bunshah, *Thin Solid Films* **107,** 21 (1983).
51. W. R. Stowell, *Thin Solid Films* **22,** 111 (1974).
52. M. Fukutomi, M. Fujitsuka, and M. Okada, *Thin Solid Films* **120,** 283 (1984).
53. B. Chapman, *in* "Glow Discharge Processes," pp. 121-123. Wiley, New York, 1980.
54. L. Holland, *Surf. Technol.* **11,** 145 (1980).
55. L. Holland, *J. Vac. Sci. Technol.* **28,** 437 (1978); **14**(C1), 5 (1977).
56. J. L. Vossen and J. J. Cuomo, *in* "Thin Film Processes" (J. Vossen and W. Kern, eds.). Academic Press, New York, 1978.
57. J. M. Poitevin and G. Lemperiere, *Thin Solid Films* **120,** 223 (1984).
58. P. Nath and R. F. Bunshah, U.S. Patent 4,336,277 (1982).
59. A. C. Raghuram, R. Nimmagadda, R. F. Bunshah, and C. N. J. Wagner, *Thin Solid Films* **20,** 187 (1974).
60. W. Grossklaus and R. F. Bunshah, *J. Vac. Sci. Technol.* **12,** *593* (*1975*).
61. J. Granier and J. Besson, *Proc. Plansee Semin., 9th 1977* (1977).

62. S. Komiya, N. Umezu, and T. Narasawa, *Thin Solid Films* **54,** 51 (1978).
63. B. Zega, M. Kornmann, and J. Amiguet, *Thin Solid Films* **54,** 57 (1977).
64. R. F. Bunshah, *New Trends Mater. Process., Am. Soc. Met.,* p. 200 (1976).
65. A. Matthews and D. Teer, *Thin Solid Films* **80,** 41 (1981).
66. Y. Murayama, *J. Vac. Sci. Technol.* **12,** 818 (1975).
67. A. Snaper, U.S. Patent 3,125,848 (1971); U.S. Patent 3,836,451 (1974).
68. A. M. Dorodnov, *Sov. Phys.—Tech. Phys. (Engl. Trans.)* **23,** 1058 (1978).
69. R. F. Bunshah and R. J. Schramm, *Thin Solid Films* **40,** 211 (1977).
70. M. Colen and R. F. Bunshah, *J. Vac. Sci. Technol.* **13,** 536 (1976).
71. H. S. Randhawa, M. D. Matthews, and R. F. Bunshah, *Thin Solid Films* **83,** 267 (1981).
72. B. Jacobson, R. Nimmagadda, and R. F. Bunshah, *Thin Solid Films* **63,** 357 (1979).
72a. F. C. Frank and B. R. Lawn, *Proc. Roy. Soc. London* **229A,** 291 (1967).
73. W. Grossklaus and R. F. Bunshah, *J. Vac. Sci. Technol.* **12,** 493 (1975).
74. R. Nimmagadda and R. F. Bunshah, *Thin Solid Films* **45,** 477, 1980.
75. V. K. Sarin, R. F. Bunshah, and R. Nimmagadda, *Thin Solid Films* **40,** 183 (1977).
76. B. Jacobson, R. Nimmagadda, and R. F. Bunshah, *Thin Solid Films* **63,** 33 (1979).
77. A. K. Suri, R. Nimmagadda, and R. F. Bunshah, *Thin Solid Films* **72,** 529 (1980).
78. K. L. Chopra, V. Agarwal, V. D. Vankar, C. Deshpandey and R. F. Bunshah, *Thin Solid Films* **126,** 307 (1986).
79. P. Lin, C. Deshpandey, H. J. Doerr, and R. F. Bunshah, in "Quarterly Progress Report" (F. A. Nichols and A. I. Michaels, eds.), TRIB-ECUT, 86-1 and 86-2, Argonne National Laboratory, Argonne, Illinois.
80. K. C. Chi, R. O. Dillon, R. F. Bunshah, S. Alterovitz, and J. A. Woollam, *Thin Solid Films* **54,** 259, 1978.
81. H. S. Randhawa, D. Brock, R. F. Bunshah, B. Basol, and O. M. Stafsudd, *Sol. Energy Mater.* **6,** 445 (1982).
82. K. C. Chi and R. F. Bunshah, *Thin Solid Films* **72,** 285 (1980).
83. B. E. Jacobson, C. Deshpandey, A. A. Karim, H. J. Doerr, and R. F. Bunshah, *Thin Solid Films* **118,** 293 (1984).
84. J. A. Thornton, *J. Vac. Sci. Technol.* **11,** 666 (1974).
85. B. Jacobson, R. Nimmagadda, and R. F. Bunshah, *Thin Solid Films* **54,** 107 (1978).
86. R. F. Bunshah, C. Deshpandey, and R. J. Lin, *in* "Quarterly Progress Report, Tribology Project" (M. Kaminsky and R. Michaels, eds.), TRIE-ECUT-84-1. Argonne Nat. Lab., Argonne, Illinois, 1984.
87. R. F. Bunshah, C. Deshpandey, and R. J. Lin, *in* "Quarterly Progress Report" (M. Kaminsky and R. Michales, eds.), TRIE-ECUT-85-2. Argonne Nat. Lab., Argonne, Illinois, 1985.
88. J. E. Sundgren, *Thin Solid Films* **128,** 21 (1985).
89. P. Agarwal, P. Nath, H. J. Doerr, R. F. Bunshah, G. Kuhlman, and C. J. Koury, *Thin Solid Films* **83,** 37 (1981).
90. P. Nath, R. F. Bunshah, B. M. Basol, and O. M. Stafsudd, *Thin Solid Films* **72,** 463 (1981).
91. H. Horose, *in* "Semiconductors and Semimetals" (J. R. Pankove, ed.), Vol. 21, p. 37. Academic Press, New York, 1984.
92. J. C. Anderson, S. Biswas, and C. S. Furtado, to be published.
93. R. C. Budhani, C. Deshandey, C. Y. Chen, and R. F. Bunshah, to be published.
94. M. S. Wittingham, *Prog. Solid St. Chem.* **12,** 41 (1978).
95. K. M. Abraham, *J. Power Sources* **7,** 1 (1981–82).
96. E. C. Zeira, M. S. Thesis, Univ. of Calif., Los Angeles (1984).

97. D. Zehnder, C. Deshpandey, B. Dunn, and R. F. Bunshah, *in* "Proc. 5th Intl. Conf. on Solid State Ionics, Part I" (J. B. Boyce, L. C. DeJognhe, and R. A. Huggins, eds.), p. 813. North-Holland, Amsterdam, 1986.
98. G. Constant and R. Feurer, *J. Less-Common Metl., Thin Solid Films* **83,** 113 (1981).
99. H. O. Pierson, *J. Composite Mater.* **9,** 228 (1975).
100. M. Sano and M. Aoki, *Thin Solid Films* **83,** 247 (1981).
101. T. Takahashi, H. Itoh, and A. Takeuchi, *J. Cryst. Growth* **47,** 245 (1979).
102. M. J. Rand and J. F. Roberts, *J. Electrochem. Soc.* **115,** 423 (1968).
103. M. Hirayama and K. Sohno, *J. Electrochem. Soc.* **122,** 1671 (1975).
104. S. P. Murarka, C. C. Chang, D. N. K. Wang, and T. E. Smith, *J. Electrochem. Soc.* **126,** 1951 (1979).
105. A. C. Adams and C. D. Capio, *J. Electrochem. Soc.* **127,** 399 (1980).
106. A. Zunger, A. Katzir, and A. Halperin, *Phys. Rev. B: Solid State* [3] **13,** 5560 (1976).
107. S. Shanfield and R. Wolfson, *J. Vac. Sci. Technol., A* [2] **1,** 323 (1983).
108. C. Weissmantel, K. Bewilogua, D. Dietrich, H. J. Erler, H. J. Hinneberg, S. Klose, W. Nowick, and G. Reisse, *Thin Solid Films* **72,** 19 (1980).
109. C. Weissmantel, *J. Vac. Sci. Technol.* **18,** 19 (1981).
110. H. Beale, U.S. Patents 4,297,387 (1981); 4,412,899 (1983), and 4,415,420 (1983).
111. R. F. Bunshah, K. L. Chopra, C. V. Deshpandey, and V. D. Vankar, U.S. Patent (1985).
112. V. V. Voskoboynikov, V. A. Gritsenko, V. M. Efimov, V. E. Lesnikovskaya, and F. L. Edelman, *Phys. Status Solidi A* **34,** 85 (1976).
113. Powder Diffraction File, Joint Committee on Powder Diffraction Data, Swarthmore, Pennsylvania (Card 15-500).
114. A. K. Suri, R. Nimmagadda, and R. F. Bunshah, *Thin Solid Films* **64,** 191 (1979).
115. T. Jamal, R. Nimmagadda, and R. F. Bunshah, *Thin Solid Films* **73,** 245 (1980).
116. M. Kodama, A. H. Shabaik, and R. F. Bunshah, *Thin Solid Films* **54,** 353 (1978).
117. M. Kodama and R. F. Bunshah, *Thin Solid Films* **96,** 53 (1978).
118. R. F. Bunshah and A. H. Shabaik, *Res./Dev.* **26,** 46 (1975).
119. R. F. Bunshah, A. H. Shabaik, R. Nimmagadda, and J. Covy, *Thin Solid Films* **45,** 443 (1977).
120. R. R. Nimmagadda, H. J. Doerr, and R. F. Bunshah, *Thin Solid Films* **84,** 303 (1981).
121. *Cutt. Tool Eng.* **34,** 4 (1982).

Ion-Beam Processing of Optical Thin Films

Ursula J. Gibson

Optical Sciences Center
University of Arizona
Tucson, Arizona 85721

I. Introduction

This article covers some of the recent developments in the use of ion beams for modifying thin films utilized in optical applications. Ion-beam sputtering, ion-assisted evaporation, and ionized-cluster-beam deposition of optical materials will be discussed. Additional references are made to studies which illuminate the mechanisms responsible for film modification. Information on this (*1*) and related topics is available in other recent reviews (*2–6*).

One of the earliest applications of high-vacuum technology was the deposition of thin metal and dielectric layers to alter the optical behavior of the substrate (*7, 8*). Thermal and electron-beam (*e*-beam) evaporation and dc and rf sputtering have made possible the production of films from almost all materials (*9*) and their utilization in a variety of multilayer designs.

The properties of a vacuum-deposited thin film of a material are rarely those of the bulk, and significant differences are observed even in films of the same material when prepared under different conditions. In early optical filter designs, these variations were not critical. However, advances in design techniques have led to specifications which may consist of a hundred layers working in concert to provide the desired spectral profile. These refined designs have made small variations of the material properties intolerable. In addition, the introduction of the laser, with its unprecedentedly high optical power densities and long-range coherence, has led to the need for improved optical thin films. Reproducible and stable optical properties, low absorptance, and mechanical durability are required for a variety of applications. Ion-beam techniques have emerged over the past few years as promising methods for producing such films.

Bombardment by energetic ions before, during, and after the deposition of thin films has been employed extensively over the past 20 years to modify the properties of the grown layers, first in discharge processes, and more recently using directed ion beams. The effectiveness of ions for film processing arises from the kinetic energy that may be imparted to them by the use of electrostatic fields and, in some cases, their chemical reactivity. Implantation of the bombarding atoms, and physical rearrangement or removal of target atoms, may occur at sufficiently high accelerating energies. The energy and the charge of the arriving ion may also stimulate chemical reactions at the surface. These processes result in changes in the optical, mechanical, and chemical properties of films deposited in the presence of ions.

Development of the process of discharge sputtering and its extension from the use of dc to rf fields made possible the deposition of thin films of a variety of materials not compatible with other techniques. In addition, it was found that films of a given material deposited by sputtering rather than evaporation tended to be more adherent, dense, and stable. In sputtering processes (*10*) an inert gas or a mixture of reactive and inert gases is ionized, and the charged particles are accelerated toward a target of the desired film material. The kinetic energy of the arriving ions is sufficient to eject atoms or molecules from the target. Some of the ejected material recondenses on a substrate to form the film. The target and the substrates are immersed in the plasma discharge in diode sputtering, and even in magnetron configurations, the substrates are bombarded by a variety of energetic particles in addition to the desired sputtered target neutrals. This bombardment, and the higher kinetic energy of sputtered versus evaporated molecules, are responsible for modification of the film properties. The bombardment flux and energy may be enhanced by combining rf and dc fields.

Since its inception, the technique of rf sputtering has been applied to the deposition of dielectric films for optical applications (*11, 12*). In the early 1970s a number of papers (*13, 14*) were published on the changes in rf-sputtered dielectric films with a simultaneously applied dc bias. This process, and ion plating, where a plasma discharge is used to bombard an evaporated film, were both observed to lead to densification of the films. Both physical and chemical changes were observed. Disruption of the usual columnar structure of rf-sputtered Corning 1720 glass, along with reduction of surface roughness, accompanied the use of a dc bias (*13*). Improved optical properties were reported by Heitmann (*15*), using ionized gases in reactive evaporation.

The pressure in the chamber during these processes was typically on the order of 10^{-3} torr, and it was pointed out by Aisenberg and Chabot (*16*) that the high density of background gas could be detrimental to the film properties. A high background pressure is usually associated with the formation of less dense films, whereas bombardment was intended to increase the packing fraction. They suggested that isolation of the ion source would permit bombardment effects at a lower ambient pressure, and that this might have significant advantages for the production of dense layers. They implemented this idea, and in 1972 (*16*) reported on the deposition of hard "diamondlike" carbon films for abrasion protection of plastics. This, and other reports, led to the development of the various ion-assisted deposition methods discussed in this review. The processes discussed here involve the combination of a relatively low background pressure ($\leq 10^{-4}$ torr) with energetic ion bombardment. This is in contrast to conventional sputtering, where the background pressure is typically 1–100 mtorr.

The application of ion-beam methods to the production of optical films began in the 1970s. Early success by Ebert (*17*), Allen (*18*), and others (*3, 19–21*) in producing low-absorptance films, combined with the commercial availability of Kaufman-type ion sources (*22*), resulted in an increase in activity in this area in the early 1980s.

Ion bombardment has emerged as a powerful technique for the production of films with desirable optical properties, from a variety of starting materials. The method is particularly beneficial in the case of dielectrics, where the conventionally deposited film is amorphous or polycrystalline due to the low mobility of the depositing atoms. For semiconducting applications, disruption of the lattice by the energetic bombardment usually leads to a film with less desirable properties. A notable exception is the formation of *a*-Si : H, where the ion beam is an easily controllable method of introducing the hydrogen. Metals have also been deposited with concurrent bombardment, and significant improvements in the adhesion and changes in the stress of the films have resulted (*14, 23–25*). In addition to bombardment during growth, ion precleaning and etching before deposition have been investigated, and postdeposition exposure has been used to modify thin optical films. The materials covered in this review include examples of these three materials categories and processing possibilities. A range of improvements in mechanical and environmental stability, changes in the real and imaginary parts of the refractive index, and control over the stoichiometry and bandgap of materials of interest have been observed.

II. Ion Sources

There is a wide variety of ways in which ion beams are used in the processing of thin films. Depending on the application, the beam may be composed of atoms of a simple inert or reactive gas, molecules of a more complex nature, or even clusters of molecules. Different source configurations are used to generate these beams, and are designed to cover a wide range of energy and current combinations.

The three sources that have been used most widely in optical thin-film work are: a hollow cathode [modified Heitmann-type source (*15*)] introduced by Ebert (*17*), the duoplasmatron (*26*), and the broad-beam Kaufman-type sources developed as space thrusters (*27*). These three sources are shown in Fig. 1. The duoplasmatron and Kaufman sources are available commercially. There are also some cold cathode sources on the market (*28*), but to date few reports in the literature on their use have appeared. A variety of designs has been used for the generation of primary ion beams, where the film is deposited directly from the beam (*29–32*). This method has been used in the deposition of both metal and dielectric films.

1. Hollow Cathode

The hollow cathode source used by Ebert (*17*) (Fig. 1a) has the advantage that it can be constructed readily in the laboratory and uses a dc discharge, requiring a single power supply. The disadvantage is that there

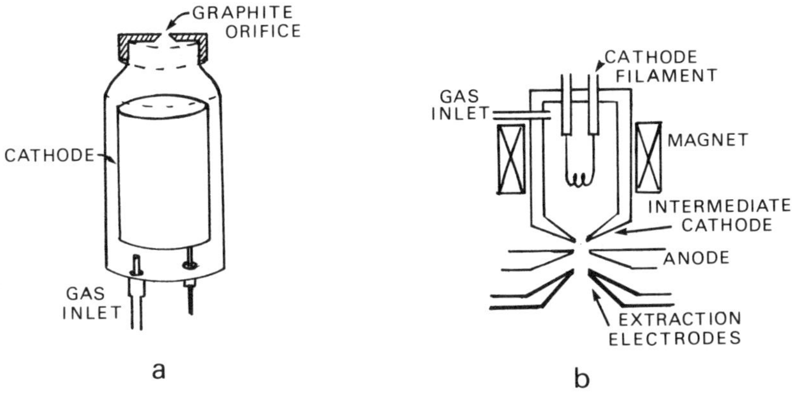

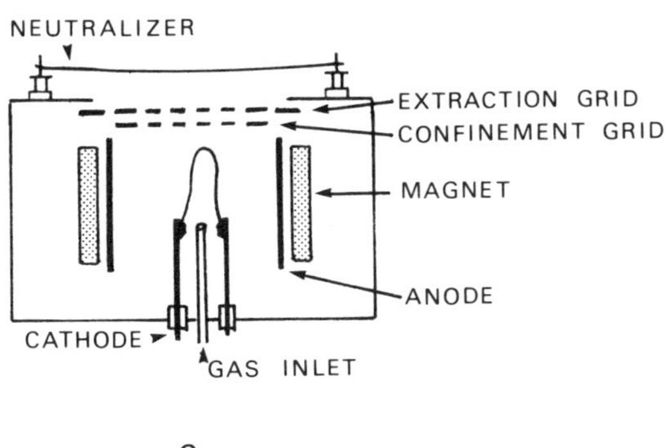

FIG. 1. Ion source configurations: (a) hollow cathode, (b) duoplasmatron, and (c) Kaufman source.

is poor control over the charged particles that emerge from the plasma chamber. The desired gas (typically oxygen) is bled into the discharge region, and a high voltage is applied between the cathode and anode. Typical power requirements are for 0.5 A at 0.5 to 1 kV. A small carbon aperture in one end of the chamber allows some of the energetic species to escape, and confines the discharge by creating a pressure differential. Little control is exercised over the emergent species, and the particles have a spread of 10 to 50 eV in energy, peaked at less than a hundred eV.

In characterization work by Allen (*18*), it was found that the flux of emitted ions was roughly linear with discharge current, and that there was a strong angular dependence of the flux. Operation at high currents resulted in rapid erosion of the anode aperture.

2. Duoplasmatron

The duoplasmatron (*26*), shown in Fig. 1b, produces a focused beam, with typical energies of 1 to 10 keV and total currents of tens to hundreds of microamps. Typical energy spreads are on the order of 10 to 50 eV. A heated filament, and a combination of electric and magnetic confining fields, create a small, high-density plasma. Since the plasma meniscus is small, the total current which can be drawn in limited by space-charge considerations. The primary advantages of this source are that the beam may be rastered electronically, and that the discharge is sufficiently isolated from the chamber that (with differential pumping) the gun may be operated with the chamber pressure in the 10^{-7} to 10^{-8} torr range. The beam size at the target is typically less than a millimeter. For many applications larger current and wider beams with lower ion energies than are accessible with this design, are desired.

3. Kaufman (Broad-Beam) Source

The Kaufman (*27*) ion source (Fig. 1c) has largely displaced the other two for research systems in recent years, due to its versatility. Using a thermionically supported discharge, magnetic confinement, and multiple apertures, extraction of large currents is possible over a wide range of energies. The energy spread of the ions in the beam is typically only a few eV, and within wide limits the beam energy and current may be varied independently. Commercially available guns can be operated with beam energies from about 10 to 2000 eV, with typical maximum beam currents of 1–2 mA/cm^2. The gun consists of a plasma chamber, which uses a cathode drop of only about 50 V, combined with thermionic emission and magnetic confinement, to sustain the discharge. The improved ionization efficiency of the electrons permits the use of relatively low system pressures ($\simeq 10^{-5}$ torr) in this configuration. Two grids are used to contain the plasma and extract ions from the discharge region. The use of multiple apertures circumvents the usual space-charge limitations, permitting large total currents. A second heated filament is used outside the discharge region to supply electrons to the emerging beam. These electrons do not typically recombine with the ions in flight, but screen the ions from one another, reducing the divergence of the beam, and eliminating charging at the substrate. The area over which the ion bombardment is uniform is a function of the size of the gun and the geometry of the system, but may be tens of inches on a side.

The coverage possible with the Kaufman source, combined with independent control over the bombardment parameters and the lower energies attainable, have made it a valuable tool in investigations of the mechanisms involved in ion-assisted deposition.

4. Primary-Beam Sources

Two configurations have been used for the generation of ion beams for direct deposition of optical films: extraction of the beam from a plasma discharge, and ionization and acceleration of an emergent evaporated beam.

The earliest reports on primary-beam deposition of optical films (*32*) were on the use of beams of ionized carbon. A magnetically enhanced rf discharge sputters carbon from the cathode, and an additional carbon extraction electrode accelerates the ions toward the substrate. Argon gas is used to sustain the discharge, and hence there is simultaneous argon bombardment of the film during growth. Differential pumping permits operation at system pressures of 10^{-4} to 10^{-6} torr, and accelerating energies are on the order of 40–100 eV. Ion sources with separate evaporation and plasma ionization chambers have also been used, and Amano (*33*)

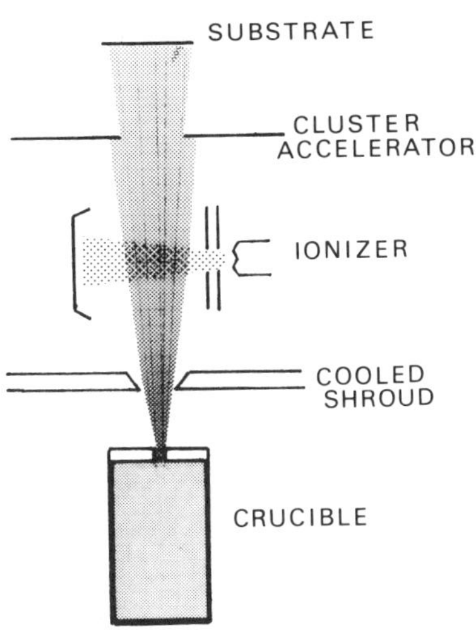

FIG. 2. Ionized-cluster-beam schematic.

used an arc discharge to generate beams of Mg and Pb with energies of 4–5 kV. Use of conventional ion lenses permitted mass separation of the beam.

In addition to the use of atomic beams, considerable interest has been generated by the description of an ionized-cluster-beam (ICB) source (*34, 35*), illustrated schematically in Fig. 2. A heated cell with a small aperture allows adiabatic expansion of the evaporated vapor, with consequent condensation of clusters of 500–2000 atoms. A narrow distribution of sizes may be obtained with appropriate conditions. The clusters are charged by electron bombardment and accelerated toward the substrate. Typical accelerating voltages are on the order of a few hundred to a few thousand volts. The large number of atoms in each cluster implies a lower energy per atom than results from other ion methods, but higher than observed for evaporation.

III. Ion–Surface Interactions

The impingement of a beam of ions on a surface may have a wide range of effects, depending primarily on the species, energy and momentum, and total flux of the beam. A number of reviews exist (*2, 5, 36, 37*), in addition to a recent book devoted to the subject (*38*). Some of the ion–surface processes will be summarized here, to provide a context for later discussion.

The energy of the arriving ion is the most important parameter in the consideration of ion–surface interactions. As the ion energy becomes comparable with the various energies of the substrate or film material, a variety of effects will be observed. Figure 3 illustrates some of the relevant energies and processes. Chemical interactions may occur between

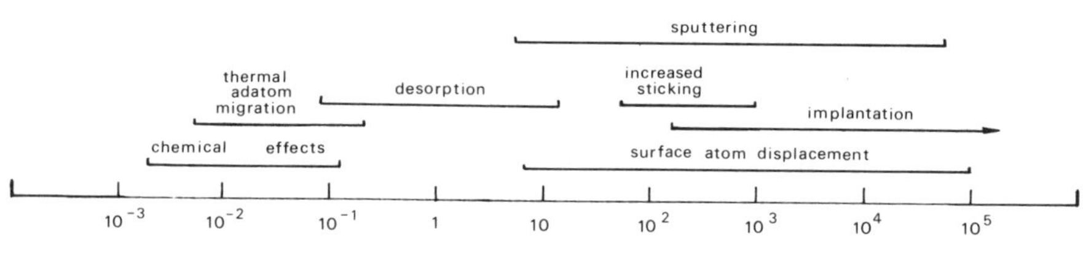

FIG. 3. Energy ranges for some of the more important processes in ion-beam-processed films [after Takagi (*36*)].

the surface and an arriving ion or atom with negligible activation energies; so that these processes are possible at all energies where the impinging ion has a finite residence time near the surface. At slightly higher energies, the surface mobility of the adatoms may be affected. The mobility of the adatoms is related to the temperature of the substrate during deposition, which is typically less than 300°C, or roughly 50 meV. For some systems, such as gold on NaCl, energies in this range are sufficient to radically alter the initial stages of growth (*39*). The kinetic energy of the arriving evaporant atoms is also relevant. The temperature of the source in a typical *e*-beam evaporation of a refractory oxide is between 1000 and 2000°C, which gives two-tenths of an eV as the upper range of the arrival energy of evaporated atoms, in comparison to the 10–1000 eV of a bombarding ion. High arrival energies may lead to removal of surface contaminants and atoms near the surface. Typical desorption energies are reported to be on the order of 1 to 10 eV (for O and CO on metal surfaces) (*40*); typical energies of sublimation for clean metal surfaces are similar (*37*). Sputtering thresholds for ejection of target atoms are less than 30 eV for most materials (*41*), with the yield rising rapidly up to an incident energy of a few hundred eV.

Sputtering is important in all ion-beam methods, whether as a precleaning method, a byproduct of the presence of ions, or a deliberate part of the film deposition process. Preferential sputtering of weakly bound material, lighter-mass, or certain chemical constituents can affect the quality of the film. The relative mass of the incident and target atoms affect the sputter yield, which can be approximated as

$$Y = K\frac{E}{\lambda}\left(\frac{M_1M_2}{(M_1 + M_2)^2}\right)$$

where K is a constant, λ is the mean free path, and the mass-dependent term times the incident energy, E, is the elastically transferred energy. A simple model for the mean free path and fitting of experimental data give

$$Y \simeq (4.2 \times 10^{-10})n_0R^2E\left(\frac{M_1M_2}{(M_1 + M_2)^2}\right) \exp\left(-10.4\frac{\sqrt{M_1}}{M_1 + M_2}E_B\right)$$

where n_0 is the number density of target atoms, R is a material-dependent collision diameter, and E_B is the heat of sublimation of the target (*37*). At energies large enough to eject surface atoms, the initial impact will be sufficient to cause a cascade of displacements. These cascade events have been modeled as thermal spikes (*42, 43*) to explain observed stress reduction and annealing effects.

For atoms arriving at the substrate with moderate kinetic energies, in addition to sputtering effects, the sticking probability of the incident rises, due to mixing implantation, and as the energy of the arriving atoms increases, more and deeper implantation occurs. The variation of the concentration of argon with bombardment current has been reported to be roughly linear at low current densities (*42*). At high energies ($\geq$2000 eV), essentially all the ions are implanted, no longer leading to sputtering, but contributing to target heating and inducing structural damage.

Kelly (*45*) has investigated the relative importance of mass and chemical bonding in determining the compositional changes in target alloys and oxides. Argon sputtering of some transition-metal oxides has been shown to lead to significant chemical reduction of the target (*46, 47*). For others, stiochiometric sputtering takes place. Kelly was able to correlate the existence of intermediate states with ΔH (heats of formation) smaller than ΔH_t (heat of atomization) with nonstoichiometric sputtering. An excerpt from his data is shown in Table I. The changes predicted by a cascade model of sputtering with these data were qualitatively correct, but less dramatic than are actually observed. He found the decomposition pressure of the materials to be a valuable predictor of the likelihood of decomposition. He also considered surface segregation, where one component preferentially diffuses to the surface. Again, more weakly bound species would be preferentially removed.

The above information has been obtained primarily in studies of the effects of bombardment on a target of known initial composition. The use of ion beams in thin-film work, by bombarding prior to or during the deposition, also affects the material significantly.

Ion precleaning of the substrate has two important effects: removal of surface impurities and generation of additional nucleation sites. Cleaning

TABLE I

HEATS OF FORMATION OF INTERMEDIATE STATES AND OF ATOMIZATION FOR SOME OXIDES[a]

Material	ΔH(eV)	Intermediate state 1	ΔH	Intermediate state 2	Total dissociation (ΔH_t)
Ta_2O_5(l)	<6.7	δ–Ta–O(s)	6.7	Ta(s)	7.1
$Ti0_2$(l)	5.1	Ti_2O_3(s)	6.4	TiO(s)	6.4
α-Al_2O_3(l)	8.0	Al(s)			6.2
SiO_2(l)	7.5	Si(l)			6.4
ZrO_2(s)	8.3	Zr(s)			7.7
ZnO(s)	6.2	Zn(s)			3.8
MgO(s)	8.9	Mg(s)			5.2

[a] After Kelly (*45*).

effects can be understood on the basis of the above discussion, but changes in growth may result from ion activation of the substrate surface. The presence of defects and bombardment with charged particles are both known to enhance the condensation process in vapor deposition (*48*), leading to films with smaller average grain size. Larger-scale changes may be induced by prolonged bombardment, with sputter removal of substrate material, which alters the optical figure of the surface (*49*). Significant changes in the nature and amount of induced roughness are also observed.

Bombardment during deposition leads to major changes in film properties. Mixing of the first few layers of the film with the substrate material may occur, which gives greatly improved adhesion. Resputtering of surface species may lead to changes in the stoichiometry and structure of the film. Since the sputter yield will be higher for low-mass atoms, preferential removal of one chemical species may occur. Resputtering of loosely bound molecules may favor the growth of denser material, as would forward sputtering of material situated above a void in the grown layer. Addition of kinetic energy by the bombarding ions enhances the mobility of adatoms (*50*) and causes a series of atomic displacements in the vicinity of the impact. Local rearrangements near the surface may either lead to a more crystalline or amorphous structure. This is discussed in some detail by Kelly (*51*). In addition, chemical decomposition of the depositing atoms may occur at the surface.

The two approaches to modeling the physical effects of low-energy bombardment are an atomistic one, using Monte Carlo methods, and the continuum thermal spike model. Some support for the spike model was found by Hirsch and Varga (*42*), who measured the current density required to reduce stress in Ge films deposited under argon bombardment to the point that substrate adhesion failure was no longer observed. They found the energy dependence to be

$$I_c \simeq 3.9 \times 10^{16}/E^{3/2} \quad \text{ions cm}^{-2}\ \text{s}^{-1}$$

which is close to the $E^{5/3}$ predicted by a thermal spike model. The magnitude of the critical current observed corresponded to the condition that the number of atoms being deposited was equal to the number that would be suffering displacement in the thermal spikes, an intuitively attractive agreement.

Recently Müller (*52, 53*) reported on Monte Carlo simulations of the growth of low-surface-mobility species under the influence of ion bombardment. He models the ion flux as having two primary effects: surface depletion due to sputtering and subsurface densification due to inward recoil (*53*). Using a Monte Carlo cascade computer and a recursive formulation, he calculates the mass density of the film. The growth of a 600 eV O^+-bombarded ZrO_2 film is modeled and displays the densification observed experimentally. However, the modeled film has more atomic oxygen incorporated than is appropriate on the basis of refractive index measurements made by Martin *et al.* (*54*) since out-diffusion effects are neglected. Müller has also performed a two-dimensional hard-disk simulation of the growth of a film under ion bombardment (*52*), using a subsurface thermal spike to simulate the energy deposition by the ions. For Ar^+ to Ni arrival ratios greater than 0.5, the model predicted a capping off of the columnar voids normally observed. This is in accordance with observations on ZrO_2 films, where bombardment during deposition led to decreased water absorption (*54*). However, the thermal spike model does not predict densification of the films, since the distribution of displacements is isotropic.

In addition to the consideration of the energy of the arriving ions or atoms, there has been some discussion of the importance of the momentum in determining the properties of films grown with ion assistance. In experiments on bombardment of Cr during deposition, Hoffman and Gaerttner (*25*) used both Ar and Xe and observed the critical flux for the relief of tensile stress. The use of two masses of bombarding ions allowed them to separate the effects of momentum and energy transfer, and they found the critical dose to be a momentum quantity.

IV. Film Properties and Analysis Methods

The use of ion processing and a multitude of other factors affect the properties of thin films, and the assessment of these properties is an important area of thin-film research. A brief review of some of the techniques which have been applied to the study of ion-beam-deposited films will be made here.

1. Optical Properties

The optical properties of a homogeneous material can be described by the optical constants n (index of refraction) and k (coefficient of extinction), related to the dielectric constant by $\sqrt{\varepsilon} = n + ik$ [note that some authors use the convention $(n - ik)$]. Knowledge of these constants, and of their dispersion (if any), is sufficient for the specification of the behavior of an ideal thin film of that material. The optical properties of the film may differ from those of the bulk, and so a variety of techniques have been developed for direct measurement of the properties of thin films. In addition to the effects of the index of refraction and absorption, the microstructure at the surface and within the film may lead to scattering of the incident light.

a. Optical Constants. In general, determination of the optical constants of a thin film of a dispersive medium is a difficult task (*55–58*). Methods include the inversion of the transmission and two reflectance measurements, made at different angles or with different polarizations. However, significant simplifications apply in the case of weakly dispersive, weakly absorbing thin films, and a determination of n and k can be made from a single spectrophotometer trace of the transmittance (*59, 60*). The interference fringe spacing gives the index, and the decay of the amplitude at short wavelengths yields a value for the extinction coefficient. The index alone may be measured accurately (*61*) by a determination of Brewster's angle for uncoated and coated portions of the substrate. A determination of the index and losses may be made in thick ($\simeq 0.5$ μm) films by using them as waveguides and measuring the coupling angles and propagation attenuation (*62*).

b. Scatter (Surface Roughness). The methods used for the determination of surface morphology are measurements of the scattered or reflected light, and imaging, either with an electron microscope or a surface profiler. A review of some of these techniques was made by Guenther *et al.* (*63*).

The total integrated scatter (TIS) gives a measure of the roughness, but no information on the autocovariance properties of the irregularities. McNeil *et al.* (*64*) describe the use of a HeNe laser and angle-resolved measurements of the scattered light to determine the surface roughness in terms of spatial frequency. Fourier decomposition of a surface profiler trace yields, in principle, the same information, although disagreement between the two methods has been reported (*65*). Direct electron microscopic examination gives a good qualitative assessment of the roughness. A scanning electron microscope (SEM) may be used for assessment of low-frequency roughness, but the resolution (typically 30–100 Å) limits the sensitivity to fine-scale structure. Transmission electron microscopy may be accomplished either by production of a shadowed replica, or by ion beam thinning or microtoming a thin slice of the film and examining the cross section directly. The resolution of the latter technique is better, since the grain size of the shadowing material (typically 20 Å) may affect replica measurements.

2. Composition (Stoichiometry)

The absorption and index of materials, in both bulk and thin-film form, are sensitive functions of the chemical composition of the material. Small changes in the ratio of metal to oxygen in refractory oxides may make large changes in both the visible and infrared properties. The use of ion bombardment during deposition is known to have significant effects on the stoichiometry, and a variety of techniques have been used to quantify these changes.

a. Infrared Absorption. An indirect assessment of the stoichiometry of some oxides may be made by measurements of their infrared spectra. Stretching and other modes of the metal–oxygen bonds may be excited optically in this region. Allen (*18*) has reported on the use of infrared measurements for the determination of stoichiometry in titania and silica films. Strong absorption in oxygen-deficient titania at 1.06 μm allowed either direct absorptance or laser calorimetric measurements, while in silica, multiple-bounce-attenuated internal-reflectance measurements were employed.

b. Rutherford Backscattering (RBS). RBS involves measurements of the number and energy of MeV α particles that are elastically backscattered toward a detector. The technique yields accurate, quantitative information on the number of atoms per square centimeter, and their mass. Careful analysis of the data also gives depth information. RBS is more sensitive to heavy elements, and is not readily applied to the analysis of thick layers, due to overlap between mass peaks. Deposition onto a low-Z substrate (C, Be) greatly facilitates the analysis. Recently (*66*), RBS has been applied to the determination of the density and porosity of thin films.

c. X-Ray Photoelectron (XPS, ESCA) and Auger Spectroscopies. There has been increasing application of surface analytical tools to the determination of the composition of optical thin films, despite the relative complexity of the required equipment. In these techniques, either an x-ray source (XPS) or an electron beam (Auger) causes ejection of electrons from the core levels of the material. The energies at which they are emitted are characteristic of the atoms, and reasonable estimates ($\pm 5\%$) of the composition may be made. The total percentage of the constituent materials in a film may not be the only relevant information, however, particularly for an inhomogeneous layer. XPS permits determination of the chemical bonding of the constituents, due to small shifts in the binding energies of the core levels as a function of the valence state. Some information on binding is available in Auger spectra also, but in general is harder to deconvolve. XPS also has the advantage that there are only small charging effects on dielectric films.

d. Other Techniques (ISS, NRA, ERD, etc.). Other ion-analysis techniques include ion-scattering spectrometry (ISS), nuclear-reaction analysis (NRA), and elastic-recoil detection (ERD). In these techniques, accelerated ions impinge on the substrate, and either the scattered ions (ISS, ERD) or the products of a nuclear reaction (NRA) are detected. All three of these techniques have good sensitivity to lighter elements. ISS is extremely surface (and thus contamination) sensitive, while the other two methods allow some depth information. A survey of these and other techniques is given by McGuire (*67*).

3. Microstructure

One of the most readily apparent differences between bulk and thin-film materials, when examined on a fine enough scale, is the microstructure. The process of nucleation, initiating the growth of the layer, results in a grain structure which leads to further modifications throughout the thickness. For growth under typical evaporation and condensation conditions, there is a fine-grained interfacial region, and the bulk of the layer has a columnar structure (where the column material may be polycrystalline or amorphous), and there are pores between the columns. The presence of energetic ions during the growth process has been seen to modify this structure, as determined by several methods.

a. X-Ray Diffraction (XRD). XRD yields information on the crystalline components of the film. A well-collimated beam of x-rays with a wavelength comparable to the interatomic spacing is incident on the sample and is Bragg reflected from any crystalline material. Analysis of the angles where constructive interference occurs gives values of the lattice spacings in the material. Preferential orientation is indicated by changes in the strengths of the peaks relative to the values published in the powder-diffraction files. Some indication of grain size and stress in the films can be gleaned from measurements of peak widths. The penetration depth of the x rays is quite large (micrometer); so the technique is best suited to the analysis of thick layers.

b. Raman Spectroscopy. When an intense beam of light is incident on a crystal, a measurable fraction of the beam may scatter inelastically off the phonons in the crystal, and emerge with a slightly shifted wavelength. The spectra are a sensitive function of the crystal structure of the material, and an adequate signal can be obtained from relatively thin layers, using an oblique-incidence technique (*68*). This technique has been used extensively for analysis of TiO_2 films, which may crystallize in one of two phases, anatase or rutile (*65, 69*).

c. Electron Microscopy. Electron microscopy, both scanning (SEM) and transmission (TEM) configurations, can be used in investigations of the microstructure of thin films. The SEM can be used for determinations of large-scale structure, and edge views of fractured films can be instructive. TEM studies of film structure are generally more informative, but sample preparation can be laborious. The penetration depth of 100 keV electrons in a typical material is only a few hundred angstroms; so either a replication or thinning technique is required. It is also possible to study thin layers deposited onto a carbon-coated grid, but the first hundred angstroms are rarely representative of the full layer. Shadow masking by oblique evaporation of Pt, followed by a transparent C coat and etching of the film from underneath, results in excellent reproduction of the structure of a fracture face (*70*), but the resolution is limited by the shadowing metal grain size. The two main thinning techniques are an ion-beam or chemical etch, or use of a microtome. The microtome is a precisely controlled knife, in common use for biological samples, which removes extremely thin sections for microscopy. Replacement of the usual glass knife with a diamond edge permits cutting of inorganic films. The slicing of the film will distort the structure, however, so a thinning technique is preferable for fine-scale information. A small-diameter rod is prepared and mechanically thinned, then the center region is etched until just perforated. The edges of the puncture will be electron transparent. Little damage is incurred with this technique, but location of the precise area of interest is difficult, and the process time consuming.

In addition to imaging capability, the TEM may be used for electron diffraction, in analogy with XRD. Samples prepared by shadowing are inappropriate for this application. Microtomed samples will yield some information, but damage due to the cutting process must be considered. A thinned sample will provide the most direct information on the crystalline nature of the film.

d. Permeation Studies. One of the least desirable changes in the optical behavior of films occurs in a humid atmosphere (*54*). In conventional films, the loosely packed structure of columns permits the adsorption of water, altering the effective index of refraction of the film. For narrowband filters this effect can be serious enough to cause a shift in the center wavelength greater than the passband of the design when the coating chamber is vented to air (*71*). The shift in the transmission curve upon venting has been used to assess the porosity of films grown by ion-assisted deposition (*54*).

Additional assessment of the porosity has been made by using the films as protective overcoats for metals placed in an etch solution (*72*). Their use as hermetic barriers for protection of optical fibers has also been reported (*73, 74*).

4. Mechanical Properties (Stress, Adhesion, Abrasion Resistance)

In addition to the improvement of optical stability effected by the disruption of columnar growth, large changes in the mechanical performance of metal and dielectric films have been observed. The measurement of adhesion and abrasion resistance is a difficult problem, and most of the results are qualitative. Measurements of stress are more quantitative, and a variety of techniques have been used. A review of mechanical properties and their determination is given by Campbell (*75*).

a. Stress. The amount of stress inherent in thin films, even those deposited at room temperature, can be enormous, corresponding to local pressures of a few kilobars. The forces are great enough that a film of a few thousand angstroms thickness may be used to induce a measureable deformation in a substrate of manageable (0.02–0.05 mm) thickness. Interferometric measurements of the deflection are made in either cantilever or disc substrates. The stress may also be determined from stress-induced birefringence (*76*), which permits measurements to be made with good spatial resolution. A qualitative determination of high stress levels may be made by observation of adhesion failure, and the failure pattern.

b. Adhesion and Abrasion. The resistance of a thin film to abrasion is a complex function of the hardness and adhesion to the substrate, and the adhesion depends on the stress, making deconvolution of the effects very difficult. However, as a practical matter, it is useful to have some measure of the behavior of thin-film coatings when exposed to mechanical abuse. Abrasion resistance is usually assessed by repeatedly dragging emery paper or an eraser across the surface. Adhesion is assessed by attempting to remove the film with adhesive tape or a small stud epoxied to the film surface. The load at which failure occurs can be measured accurately for the stud system, if the epoxy adheres sufficiently well. Dragging a stylus across the film, while linearly increasing the stylus loading, has also been reported (*76a*).

V. Substrate Bombardment

Glow-discharge cleaning of substrates, by exposure to a dc diode plasma, has been employed for a number of years in thin-film work (*77*). Ion bombardment, combined with heating of the substrate, has been a standard technique for a number of years for production of atomically clean surfaces in ultrahigh-vacuum applications. Recently, there has been an increase in the use of ion beams in conventional vacuum systems for the precleaning or modification of the substrates before thin-film deposition. The primary observed effects are (1) a change in the surface morphology, and (2) improved adhesion.

An ion beam may actually be used to machine optical components (*49, 78, 79*). The use of a rotating mask or focused directional beam permits the production of aspheres. Concurrent with the erosion, there may also be increases in the surface roughness of the optic. Extreme roughness, formed by etching of MeV ion tracks, has been used to produce antireflection (AR) layers on plastics, due to the resulting gradient of the index of refraction (*74*).

Changes in the surface roughness of the substrate may be induced by preferential sputtering from different crystal planes or locally different stoichiometries or by shadowing effects due to impurities on the surface (*38*). Improved adhesion may result from an increased density of nucleation sites, allowing improved accomodation of strain at the surface, activation of a surface layer for improved chemical bonding, or alterations in the surface structure and chemical composition.

Several workers have reported on the change in the surface microstructure under prolonged ion bombardment (*38, 49, 80, 81*). These effects have been studied by measurements of the scattering characteristics, and by direct electron microscopic observation. The changes observed are dependent on the energy, angle, duration, and current density of the bombardment. In general, removal of more than a few thousand angstroms results in increased roughness. The scale of the resulting roughness is strongly dependent on the bombarding energy, with lower energies ($\simeq$4 versus 12 keV) resulting in a finer structure on BK-7 glass (*49*). In optical studies of substrates eroded at 500–1000 eV beam energies, Hermann and McNeil (*23*) were able to remove a half-micrometer of BK-7 with no observable increase in scatter. On other amorphous substrates (vitreous silica and Dynasil), they observed doubling of the surface roughness at $\lambda = 0.63$ μm on removal of similar amounts. Greater increases in roughness have been reported for ZnS, Cu, Si, Mo, and ThF_4 (*80*).

Johnson and Ingersoll (*81*) have reported an interesting technique for ion polishing, using a photoresist overlayer on fused silica. The rough surface is covered by a spun coating of 0.5 μm of the polymer. This results in an outer surface which is extremely smooth. By correct choice of incident angle and bombarding species, the erosion rates for the polymer and substrate material may be matched. Milling is performed until all of the polymer (and all of the high regions of the substrate) are removed. Scanning electron micrographs demonstrated the removal of scratches left after polishing with a 1 μm grit.

Ion bombardment of the substrate followed by deposition also results in changes in the properties of the grown film. This technique has been used to increase the adhesion of Ag (*23*), Ga–As–Se (*82*), and ZnS (*83*). An increase in the density of nucleation sites and removal of sharp atomic steps in cleaved NaCl has also been reported (*50*). Ion bombardment prior to deposition increased the etch resistance of metal films overcoated with SiO_2 and ZrO_2 prepared by ion-assisted deposition (*72*).

VI. Postdeposition Bombardment

Ion beams, incident on the substrate after the deposition of a layer, can be used either to modify or to remove film material. The latter application has received a great deal of attention in the semiconductor industry, and will be mentioned here only briefly. The effects of bombardment after deposition are limited by the range of the ions in the target material. The value of the mean penetration depth for Ar→Si at 10 keV is only 120 Å (*84*), and this increases to a few hundred Å at 100 keV. Thus, postdeposition treatments have typically involved higher-energy ions.

Ion bombardment has been reported to alter the crystallinity of materials, either driving them toward a more amorphous or more crystalline state (*51*). This effect has been observed in neutron-irradiated silica (*85*) and in ZrO_2 (*43*), which crystallized under bombardment by a 50 keV Kr^+ beam. Saxe *et al.* (*71*) looked at water absorption by ZrO_2/SiO_2 and TiO_2/SiO_2 narrow-band filters, which were bombarded after deposition, but before exposure to atmosphere. For 3 keV Ar^+ bombardment, they found that the stability of the TiO_2 filters was improved, while that of the ZrO_2 was not. It was suggested that the increased disorder induced by ion bombardment in TiO_2 led to an increase of the volume of the material near the surface, plugging the columnar pores that are normally observed in thin films of this material. In ZrO_2, bombardment leads to an increase in crystallinity, which might lead to densification of the material near the surface, opening the pores still further.

Extreme densification of film material has been observed by Chopra *et al.* (*86*) for obliquely deposited $Se_{0.75}Ge_{0.25}$ films bombarded with 50 keV He^+. They observed a thickness contraction of 39%, with an associated increase of the index of nearly 10%, and a shift in the band edge. They were able to use this bombardment for the replication of fine-line gratings.

Ion milling has been used both for the generation of structures for integrated optics (*87, 88*) and as a diagnostic for the composition of multilayer films (*89–91*). Hermann and McNeil (*90*) reported on the use of optical monitoring of the reflectance of a multilayer film during sputtering. Comparisons of the measured-to-calculated reflectivity curves made possible qualitative determinations of the thicknesses of the layers. In a similar experiment, in transmission, Macleod (*89*) and co-workers were able to detect the presence of a ≈20 Å layer of metal in a coating of unknown composition. While not as sensitive or powerful a technique as Auger profiling, use of a Kaufman source has the advantage of speed and compatibility with dielectric materials. Profiling optical layers with x-ray photoelectron spectroscopy has not been practicable, due to the thicknesses involved and the large area of the imaging spot, requiring rastering of the ion beam. The new small-spot XPS equipment should improve the situation.

VII. Ion-Beam Deposition of Optical Films

The use of ion beams before and after the deposition process may have significant effects on the optical response of the film–substrate system. However, much larger effects will result if the ion beam is used as part of the deposition process, either as a direct source of the film material, a source of ions to sputter a target (with subsequent deposition of the sputtered material), or as a source of either inert or reactive ions, bombarding the film during deposition by another method. These three techniques are known as primary-beam deposition, ion-beam sputtering (IBS), and ion-assisted deposition (IAD). In this section, a review of results obtained by these three methods for the deposition of optical films is presented, grouped by material. Related studies have also been presented by Martin (*1*).

1. Dielectrics

Dielectric materials, having a bandgap beyond the visible, are used extensively for multilayer wavelength-selective coatings. The index difference between layers changes the reflectance and any absorption is undesirable. In a stoichiometric deposit, there will be no efficient mechanism for absorption, and all of the light will be either transmitted or reflected. For high optical power levels, this has the obvious advantage that less damage will be incurred, and less cooling required. Of these materials, those with especially high or low refractive index are particularly desirable, since the index difference between adjacent layers determines the number of quarter-waves required to achieve effective antireflection or reflection enhancement. High-refractive-index materials are particularly difficult to work with, since there is a roughly inverse relationship between the index and bandgap of nonmetallic materials. This

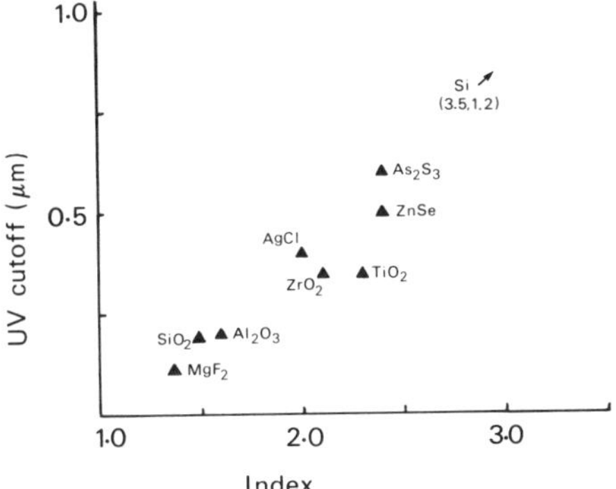

FIG. 4. UV cutoff versus index of refraction for transparent materials used in optical coatings.

effect is seen in Fig. 4, in which is a plot of the ultraviolet cutoff of several materials versus an approximate index, chosen in the region of low dispersion (typically 0.8 μm). Defects and impurities tend to smear the band edge and bring it into the visible, leading to absorption. Oxides of the transition metals, such as Ta or Ti, are used extensively because of their high index, but the unfilled *d* shells permit a multitude of oxidation states, not all of which are transparent. This makes the materials particularly sensitive to deposition conditions.

In addition to the optical properties of the films, the mechanical and morphological properties are of considerable interest because of their influence on the stability and durability of the coatings. Microstructure on the scale of the illuminating radiation will introduce undesirable scatter.

The combination of so many requirements on the properties of dielectric films has led to the use of many techniques for their deposition and analysis, including ion-beam processing.

a. Titanium Dioxide (*Titania*, *TiO_2*). Titanium dioxide, in bulk, has three stable crystalline phases: anatase, brookite, and rutile. Of these three, rutile is the only transparent material, and is strongly birefringent, with refractive indices of 2.9 and 2.6. Anatase is less birefringent, and has an index of about 2.5. The brookite form is rarely observed in thin films. In addition to the different crystal structures, TiO and Ti_2O_3 are stable, but absorbing, materials. The high index of TiO_2, combined with desirable mechanical properties, have made it the subject of much thin-film research despite these complicating factors. Promising results have been achieved for the production of films with both IBS and IAD. For IAD, heated substrates and low fluxes at moderately high energies give the best optical quality.

(*1*) *Optical properties*. Titania was one of the first materials studied using ion sources, due to the difficulties in depositing stoichiometric films. Reactive evaporation at high enough pressures to encourage stoichiometry resulted in fragile films (*92*). Early studies using a Heitmann-type cold cathode source (*15, 18, 93*) demonstrated that an increase in index and decrease in absorption could be obtained by irradiation of a thermally evaporated film with low-energy oxygen ions. The mean energy of the ions in these experiments was about 50 eV, and in the work of Ebert (*17*) the ratio of arriving ions to atoms that yielded the best films was about 10 to 1. Both Küster (*93*) and Ebert (*17*) reported that a slight improvement in the films was obtained with Ti_2O_3 as a starting material, rather than TiO, and found less absorption for films deposited without substrate heating. Ebert studied the effect of the bombarding species and found that neutral excited oxygen decreased the absorption, but that negatively charged ions resulted in higher indices and lower absorptances. For unheated substrates, irradiation with 0.3 mA/cm^2, and negative bombardment, he found an index of 2.3 at $\lambda = 514$ nm, with an absorption coefficient of 50 cm^{-1}. With concurrent substrate heating, he found $n = 2.4$ at 550 nm. In similar work, using only TiO as the start material, Allen (*18*) reported a maximum index of ≈ 2.43 at 514 nm, with an absorptance of 30 cm^{-1}. He reported an optimum beam current, beyond which there was no increase in index, and a slight increase in the absorption.

Oxygen-ion-assisted deposition of titania has also been studied using Kaufman sources. High-energy bombardment affects both the stoichiometry and packing density of the films, whereas the primary effect of low-energy bombardment seems to be on the stoichiometry. No significant difference in absorption was found between films deposited from TiO and TiO_2 start materials, when low-energy (60 eV) bombardment was used (*92*). At ambient temperature with a 30 eV beam, only a slight increase of the index was observed, but the absorptance was decreased, showing a minimum value with a bombarding current density of about 150 $\mu A/cm^2$ (*94*). Allen (*95*) reported on the deposition of films onto 200°C substrates with Ar, Ar + O_2, and O_2 beams. He obtained the best results with O_2 bombardment at moderate (300 eV) energies. He found an increase in the index with an ion-to-molecule arrival ratio up to about 0.12, and obtained a maximum value of $n = 2.5$ at 550 nm. Use of ion energies above 300 eV with constant current resulted in a decrease of the index. Similar results were reported by McNeil *et al.* (*92*). An index of 2.5 in the visible, with no measureable increase in k, was observed with a substrate temperature of 250°C. The bombarding conditions were similar, with 15 $\mu A/cm^2$ of positive ions at an energy of 500 eV. Martin *et al.* (*54*) reported similar index results for ambient substrate temperature when high-energy bombardment was used. With 600 eV and 16 $\mu A/cm^2$, they reported an index of 2.52 and $k = 8 \times 10^{-3}$. Increasing the energy to 750 eV and the current density to 48 $\mu A/cm^2$, they observed a slight decrease in the index, but a decrease in k of more than an order of magnitude.

Early work by Chopra and Randlett (*96*) demonstrated the feasibility of rapid ion-beam sputtering of TiO, but it was not until recently that IBS was applied to the problem of TiO_2 deposition (*46, 47, 69, 97–101*). Films were deposited by bombarding a metallic target with an argon or argon–oxygen beam at high energies (800–1200 eV). The sputtered material was then collected on the substrate. In some cases, the growing film was also bombarded with low-energy argon to modify the stress.

IBS also results in the deposition of coatings with an increased index. Values over 2.5 have been reported in the visible, at 530 nm (*101*), and at 633 nm (*46*). Visible absorptance (20–50 cm^{-1}) comparable to that obtained with some of the ion-assisted evaporation was reported. Tests of the performance of these films at 1.06 μm have also been made. An index of 2.27 with a $\lambda/2$ absorption of 3.8×10^{-4} (*99*) and an index of 2.42 with $\lambda/2$ absorptance of 8.5×10^{-5} have been reported. Laser damage thresholds are moderately high but have not yet exceeded those observed in rf-sputtered TiO_2 (*97*).

Lower levels of scatter have been observed from IAD and IBS titania than are observed from evaporated films (*65, 69, 92*), and the scattering has been correlated with changes in the crystal structure of the films. However, while Al-Jumaily *et al.* (*65*) correlated smoothing with an increase in anatase Raman scattering in the films, Hsu *et al.* (*69*) saw an increase of two orders of magnitude in the elastically scattered light when IBS films were thermally annealed to enhance their crystallinity.

(2) *Composition.* The change in composition in titania films has been qualitatively assessed by measurements of the absorption at 1.06 μm (*17, 18*). More quantitative studies have been made with ion scattering techniques and x-ray photoelectron spectroscopy. Films deposited by IAD (*92*) showed contamination levels of 0.01–0.05% W from the filaments used, and varying amounts of H and C. Typical impurity levels were 2–5% of H, and $\leq 1\%$ of carbon. The tungsten contamination correlated directly with the bombarding current, whereas the H and C content seemed more dependent on the substrate temperature and bombarding energy. The H content was slightly lower for high substrate temperature and 500 eV bombardment, which correlates with observations of reduced porosity and water adsorption. The carbon levels were significantly increased in films deposited onto hot substrates, and somewhat increased

for high bombardment fluxes, particularly for the 500 eV ions, where carbon extraction grids are used.

The ratio of Ti to O in films deposited at 250°C was seen to increase from 1.85 to 1.91 with bombardment (*102*). Higher currents ($\approx$180 μA/cm^2) were required to induce this change at 60 eV than at 500 eV (9 μA/cm^2). Higher currents at 500 eV led to an increase in the oxygen-to-titanium ratio to 1.924. The stoichiometry changes do not correspond directly with either the energy delivered (twice as high for 60 eV) or the momentum (10 times as high for the 60 eV case). The activation energy for the chemical reaction presumably dominates.

The stoichiometry and bonding in TiO_x films deposited by ion-beam sputtering has been studied in detail by XPS (*46, 47*). Analysis of the peak position and shape of the Ti 2*p* lines gives an indication of the bonding states of the metal ions. Demiryont and Sites (*46*) have reported a mixture of stoichiometric and substoichiometric material, with a ratio dependent on the oxygen percentage in the sputtering beam. They found no impurities in the films at the 0.5% level, although there was a hint of a peak at the Ar photoelectron energy. Their studies also demonstrated the extreme sensitivity of the stoichiometry to bombardment; large changes were observed in the Ti signal during sputter cleaning of the sample, if carried on too long. In related studies, it was observed that the oxygen peak did not shift during sputtering (*47*), indicating that dangling Ti, rather than O, bonds are responsible for the absorption.

(*3*) *Microstructure*. Electron-beam-evaporated TiO_2 films deposited onto room-temperature substrates are essentially amorphous (*65, 69*), showing no crystalline peaks by either XRD or Raman spectroscopy. In early studies of ion-assisted deposition, with low-energy oxygen bombardment, Allen (*18*) observed no measureable crystalline structure. Later studies (*65*), using Raman spectroscopy and with higher bombarding energy (500 eV), showed an increased presence of anatase crystallites in IAD films. Simultaneous heating and bombardment of the film led to even further crystallization. Hsu *et al.* (*69*) performed Raman studies on films deposited by *e*-beam and IBS processes, and found that while the as-deposited films had similar response, annealing led to the preferential formation of either anatase, rutile, or a mixture of crystallites, depending on the deposition conditions. Small seed crystals, whose structure depended on the deposition conditions, grew during the annealing. IBS films deposited with a high current density in the sputtering beam showed the greatest rutile character. Annealing also led to an increase in the index, but an accompanying increase in scattering and the extinction coefficient.

Measurements of the optical fringe shifts on exposure to air (*54*) indicated that there was a reduction in the porosity of titania films. Films deposited with 600 eV bombardment showed no measurable shift, while those with either heated or ambient substrates had shifts (2%, 9%) indicative of the uptake of significant amounts of water vapor.

(*4*) *Mechanical properties*. Allen (*18*) reported on the stress of reactively evaporated and ion-bombarded titania. He observed tensile stresses for all of the films, with an increase in magnitude of the stress for ion-assisted depositions. McNeil *et al.* (*94*) noted no effect on the stress, using a Kaufman source at comparable energies. IBS-deposited films were reported to have compressive stress (*99*), and an adhesion strength in excess of 10^4 psi, as measured by a pull-tester.

b. Silicon Dioxide (*Silica*, SiO_2). SiO_2 has a number of transparent, naturally occurring mineral forms, with indices ranging from 1.41 for the amorphous hydrated opal form to 1.55 for crystalline quartz. SiO also occurs but is not transparent. Fused quartz, commonly used for substrates, is amorphous and has an index of 1.46 in the visible. Vacuum-deposited films are also amorphous, with a similar value for n. SiO_2, due to its low index, is commonly used in multilayer designs with TiO_2 or other high-index materials. For this reason, many of the early studies on titania were complemented by investigations of silica. The index of refraction varies little with SiO_2 films; so the bulk of the reports have been on the level of absorption. The best results for IAD have been obtained with low-energy bombardment, and for IBS using an SiO_2 target.

(*1*) *Optical properties*. Heitmann (*15*) investigated reactive evaporation of SiO with oxygen in the chamber, and found that unless a discharge was struck so that oxygen ions were incident on the substrate, poor films resulted. With ions present, he was not able to distinguish between the transmission of the substrate and the substrate–film system. Ebert (*17*) used a similar source to deposit fully oxidized films onto a heated substrate, achieving an index of 1.46, again with no observable absorption. These studies were followed by a more intensive study of the absorptance behavior by Allen (*18*). On unheated substrates, he reported a decrease in the absorption at 1.06 μm by a factor of six with increased bombardment current. Increasing the substrate temperature increased the absorption slightly up to 200°C, and doubled it by 325°C.

Studies with a Kaufman-type source onto unheated substrates indicated that optimum bombardment conditions exist for each energy, and that low-energy bombardment may be superior for this material (*94*). McNeil *et al.* measured the change in transmission of an SiO_2 substrate upon the application of 0.4 μm layers, deposited with varying levels of bombardment. 30 eV bombardment resulted in a higher transmission at all values of the current density, with a peak improvement at 60 $\mu A/cm^2$. For

500 eV oxygen bombardment, there was an improvement over unbombarded films, but at all current densities the increase in transmission was less than that for low-energy bombarded samples. Substrate temperature was seen to have little effect. Argon bombardment of SiO_2 was also investigated (*54*) and, not surprisingly, increased the absorption. When an oxygen backfill was used in conjunction with the Ar^+, the absorption decreased again, and the stability of the layer was improved.

Ion-beam-sputtered films of SiO_2 also display decreased absorption when an argon–oxygen mixture is used. Sites *et al.* (*101*) reported slightly higher indices (1.49 at 0.53 μm and 1.47 at 1.06 μm), and low absorption levels ($<10^{-4}$ at 1.06 μm) in films produced by sputtering of a pure silicon target with a reactive ion beam. They found that similar films could be deposited with a silicon dioxide target with 10% oxygen in the beam rather than 25%. Ionized-cluster-beam deposition resulted in films of SiO_2 with an index of 1.46 and good step coverage (*103*).

Scatter measurements were also made on IAD silica films, and lower levels were observed at all spatial frequencies (*65, 92*) than for conventionally deposited films.

(*2*) *Composition.* For silica, as for most oxides, excess absorption is usually related to oxygen deficiency. Optical and other methods have been applied to the determination of the stoichiometry of silica films. Using attenuated total reflectance to increase the IR absorption signal, Allen (*18*) was able to characterize the bond lengths in films deposited with and without ion bombardment. The bonding is different for the different oxidation states, allowing a determination of the composition. He found that the films deposited with IAD had a composition essentially that of SiO_2, and lower levels of both water and hydrogen compensated silicon bonds. A similar shift in the stoichiometry was found by Dudonis and Pranevicius (*21*), using 5 keV oxygen bombardment during growth. Elastic-recoil detection studies of similarly deposited films (*94*) did not indicate such a large reduction in H concentration. At 500 eV, with current densities about 100 $\mu A/cm^2$, the H content was smaller, but these conditions result in absorbing films.

XPS has been used in the study of IBS-deposited SiO_2 films (*47, 99, 100*). Those deposited from a silicon target were found always to be somewhat deficient in oxygen, even with a 60% oxygen sputtering beam. With an SiO_2 target, the films were stoichiometric at a 10% oxygen level in the beam (*100*). More detailed studies (*47*) revealed that the linewidth of the oxygen 1*s* peak was greater than that observed for evaporated films. It was proposed that this was due to a greater randomization of the bond lengths and angles, and thus related to the greater absorption in these films.

(*3*) *Microstructure*. Along with the changes in stoichiometry, there are changes in the morphology of these films. Since both the evaporated and ion-beam-deposited films are amorphous as determined by XRD (*18, 54*), the changes are only apparent on a larger scale. Reduced porosity has been observed in films deposited with high-energy (>500 eV) bombardment, as determined by both water adsorption after deposition (*54*) and improved protection of metallic underlayers (*72*). A dramatic increase in the durability of the overcoated metals was observed, with the best results obtained with films where ion bombardment was used throughout the process, including precleaning, bombarding the metal layer, and bombarding the dielectric overcoat.

Cole *et al.* (*104*) compared the water absorption in films of SiO_2 deposited by *e*-beam evaporation, ion-assisted deposition, and IBS. In measurements of the depth of the IR absorption band after soaking in H_2O for 30 min, they found much less water uptake by 300 eV IAD films than those deposited with no bombardment, and essentially none by the IBS films. It is possible that IAD films deposited with higher-energy bombardment would result in comparable performance.

(*4*) *Mechanical properties*. Stress levels have also been measured in SiO_2 films. Both Allen (*18*) and McNeil *et al.* (*94*) found a slight reduction in compressive stress for IAD films. Significant reduction of the compressive stress of IBS films was found when the growing film was bombarded with a second ion source (*99*). The reduction was greater for greater bombarding energies. Good adhesion of these coatings was also observed.

c. Zirconium Dioxide (*Zirconia*, ZrO_2). Zirconium dioxide is another high-index material which has a variety of naturally occurring crystalline phases. There are monoclinic phases stable at room temperature with indices of 2.1–2.2 in bulk form. At higher temperatures, a cubic phase is found. In almost all zirconia, there is some inclusion of Hf. The index of a film deposited onto a room-temperature substrate is about 1.9, and onto a hot substrate, 1.95–1.80, varying with the thickness of the layer (*54*). Extensive research on the effects of IAD on the properties of this material has been conducted (*17, 43, 54, 72, 105–108*). The best results are obtained for heated substrates, moderately high bombarding energies, and rather low fluxes.

(*1*) *Optical properties*. An early report by Ebert (*17*) noted a decrease in the absorption of ZrO_2 films deposited by IAD with low-energy bombardment, but more work has been done at energies of a few hundred eV per ion. The effects of argon bombardment, argon bombardment with an oxygen backfill, and direct oxygen bombardment have been assessed. In all cases, there was an increase in the index of refraction, but without oxygen present the absorption also increased. For 600 eV, 16 $\mu A/cm^2$ Ar with an oxygen backfill to a partial pressure of 5×10^{-3}, Martin *et al.* (*54*) measured an index of 2.05 at 550 nm, with an extinction coefficient of less than 7×10^{-4}. This absorption was comparable to that obtained by evaporation onto a heated substrate, and a factor of two less than that obtained for Ar bombardment with no oxygen present. The variation of index with bombardment current for Ar (*106*) showed a sharp peak of 2.15 with a current density of 40 $\mu A/cm^2$, with a falloff for higher values. Heating of the substrate increased the peak index to 2.23, which is greater than that of the bulk. Oxygen bombardment led to a rapid increase in the index up to 75 $\mu A/cm^2$, and gradual saturation for higher currents at a value of about 2.19. Heating the substrate while bombarding with oxygen also yielded an index of 2.23 (200 $\mu A/cm^2$). Measurements of waveguide losses (*105*), which are dependent on both the scattering at the interfaces and the absorption in the film, showed greatly improved performance for IAD films. The losses for an unheated, evaporated film were greater than 10 dB/cm, while those for IAD films were 4 and 2.1 dB/cm for films with 100 and 1200 eV oxygen bombardment, respectively. The index for the higher-energy bombardment was slightly increased.

(*2*) *Composition*. The composition of ZrO_2 films has been assessed using RBS, XPS, and NRA. The argon content of the bombarded films was seen to rise to about 1.5% as the current density was increased to 50 $\mu A/cm^2$. It is interesting to note that the sharp rise coincides with the peak index observed for Ar bombardment. RBS also indicated the presence of small amounts of Hf (*108*), resulting from the presence of that impurity in the starting material for evaporation. The argon concentration was reported to be less for films deposited onto a heated substrate. NRA was used to detect H in the films, indicative of the adsorption of water into a porous microstructure.

(*3*) *Microstructure*. The form of these films, both on the atomic and larger scales, has been investigated by a number of techniques. The crystallographic complexity of the material has encouraged the extensive use of XRD. Ion bombardment of the growing film leads to formation of an fcc phase (*106*), which is not observed in either ambient or hot-deposited films. The ambient deposited films are amorphous, and heated substrates produce monoclinic crystallites. Heating and bombarding simultaneously produced a mixture of monoclinic and cubic phases (*106*).

Studies of the gross morphology have been made with both scanning electron microscopy and, indirectly, by studies of the porosity of the films. Protection of metal films from an etchant solution by an overcoat of zirconia was greatly improved by the use of IAD (*72*). A dramatic increase in the stability of ZrO_2 layers on exposure to a moisture-containing atmosphere was correlated with the disruption of the usual columnar structure of the films (*54*). Nuclear-reaction analysis was used to measure the hydrogen profile in films deposited under different conditions, to give a more quantitative measure of the uptake of water by film pores (*106*). For films deposited onto heated substrates, they found that the H concentration in the IAD films was low, and constant throughout the film. For conventionally deposited films, they found high concentrations at the surface of the film, decreasing toward the substrate, in agreement with the porous-film model. For unheated substrates, the effect of IAD was less dramatic (*108*). The film density, determined by weighing and Talystep measurement of the thickness, was also seen to rise with ion bombardment (*106*).

d. Magnesium Fluoride (*MgF_2*). Magnesium fluoride is a tetragonal crystal with slight birefringence, which is insoluble in water and alcohol, but can be dissolved by nitric acid. It has one of the lowest indices of refraction of the inorganic compounds compatible with vacuum deposition, and is transparent from less than 0.2 to about 6.5 μm. Approximately stoichiometric deposits result from resistive or *e*-beam evaporation, but an elevated substrate temperature is required to produce a dense, mechanically stable film. Sputter deposition of the compound has not been successful, due to preferential removal of the fluorine, yielding absorbing films. Ion-assisted deposition of MgF_2 results in durable films with good optical properties, if there is sufficient oxygen present to saturate dangling bonds left by fluorine depletion.

(*1*) *Optical properties*. Early attempts to deposit MgF_2 by ion-assisted deposition resulted in durable films on room-temperature substrates (*23*) with little change in the refractive index, but had 0.5% absorption in the visible for a layer 0.13 μm thick. The bombardment was performed with 700 eV Ar^+ and a current density of 50 $\mu A/cm^2$. Using low-energy (30–250 eV) Ar^+, Kennemore and Gibson (*109–111*) reported the deposition of 0.2 μm layers with no detectable visible absorption ($<0.1\%$), and an index essentially that of a film deposited onto a heated substrate. The ultraviolet absorption edge shifted to slightly longer wavelengths (270) nm. Brewster's-angle measurements of the index at 0.63 μm yielded values of 1.38 to 1.4 (*111*). Simultaneous heating and bombardment of the films resulted in strongly absorbing films (*112*). Recently, Martin and Netterfield (*113*) reported on 0.4 μm thick films of MgF_2 deposited with water and oxygen bombardment at higher energies (700 eV). They found indices at 550 nm of 1.35, 1.39, and 1.40 for deposition onto unheated substrates with no ions, oxygen ions, and H_2O ions as the bombarding species. The absorption in the films was reported to be less than 0.1%. Using Ar^+, they reported an index of 1.41, but an absorption of 1.5%. Bombardment with halocarbons at intermediate energies has also been used (*114*), and results in low-absorption films.

Ionized-cluster-beam deposition of MgF_2 with an accelerating energy of 4 kV resulted in films with an index of 1.38 (*115*) and low scatter losses.

(*2*) *Composition.* The stoichiometry of IAD magnesium fluoride has been assessed by RBS and XPS. In all cases the films were seen to have some fluorine deficiency, which increased as higher bombarding energies (*110*). The fluorine-to-magnesium ratio, as determined by RBS, dropped from 1.93 for a thermally evaporated film to 1.86 for one bombarded with 150 eV Ar^+ during growth. The ratio for a film deposited with halocarbon bombardment was 1.96–1.98 (*114*).

Impurities in the films included small amounts of Ta and W from the evaporation boat and the neutralizer filament, oxygen, components of the bombarding species, and a surface layer of carbon. Argon inclusion increases with bombarding current (*112*) and energy. For argon bombardment at 150 eV and 30 $\mu A/cm^2$, about 1% Ar incorporation was reported (*110*), but the oxygen content rose from less than a percent to nearly 10% in the IAD film. Residual water vapor in the system was the most likely source for the oxygen. A correlation between fluorine deficiency and oxygen uptake suggested substitution of oxygen at dangling magnesium bond sites (*109*). The work of Martin and Netterfield (*113*) on O_2 and H_2O bombardment supports the hypothesis that oxygen is important for the production of nonabsorbing films. Films deposited with 300 eV Ar bombardment had lower oxygen concentrations and were absorbing (*111*). Preliminary measurements of the carbon content of Freon-bombarded films indicate 5% inclusion, but some of this may be attributable to the C substrate.

(*3*) *Microstructure.* The microstructure of ion-beam-deposited MgF_2 has been investigated using x-ray diffraction and transmission electron microscopy. Ionized-cluster-beam-deposited films on ambient-temperature substrates were polycrystalline, but the usual preferential crystal orientation was not observed (*115*). Films grown by IAD were reported to be amorphous, as determined by both XRD and TEM diffraction studies (*109*).

The change in index of the films on exposure to air, an indication of porosity, was reported by Martin and Netterfield (*113*) to be eliminated for 700 eV bombardment when the ion-to-molecule arrival ratio was 0.16.

(*4*) *Mechanical properties.* The increased energy associated with ion processes improves the mechanical properties of MgF_2 films deposited by either ICB or IAD. Reduced stress has been observed in IAD MgF_2 films (*23, 111, 116*), as indicated by the reduction of stress-induced crazing.

Deposition of layers of 1.6 μm thickness, which comprise quarter-waves at the long-wavelength transmission edge of MgF_2 are possible (*44*). Measurements of the stress-induced birefringence in films deposited on annealed BK-7 support the qualitative observations (*116*).

Takagi and Yamada reported improved abrasion resistance and adhesion of films deposited on plastics and glass by ICB (*115*). Qualitative results on the adhesion and abrasion resistance of IAD films on room-temperature plastic and quartz substrates indicated great improvements in their durability (*23, 109–111*). Subsequent tests with a dynamically loaded scratch tester (*116*) proved the adhesion of these films to be superior to that of films deposited conventionally onto a heated substrate.

d. Tantalum Pentoxide (Ta_2O_5). Tantalum pentoxide is a rhombohedral crystal which is largely insoluble in acids and has an index over 2.0 when prepared in thin-film form. There exists a hydrated gel form which is less resistant to chemical attack, and it can also be reduced to TaO_2 (Ta_2O_4), which is an absorbing material. Considerable effort has been made to characterize IBS films of the material (*47, 99–101, 117*), and it has also been deposited by IAD (*17, 105, 118, 119*).

(*1*) *Optical properties*. Ebert (*17*) reported the use of IAD to reduce absorption in reactively evaporated tantalum pentoxide films, and recently there has been further investigation into the material. McNally *et al.* (*119*) reported on IAD films deposited onto heated substrates. With low-energy oxygen bombardment, the index at $\lambda = 350$ nm rose slowly from 2.15 to a maximum of 2.25 with a flux of about 60 $\mu A/cm^2$. For 500 eV ions, there was a sharp maximum in the index with current densities below 10 $\mu A/cm^2$, and a subsequent decrease. Using a heated substrate and low current densities, they were able to deposit films with low absorption levels. Current densities above 20 $\mu A/cm^2$ resulted in absorbing films. Higher energies (1200 eV) were used in the production of waveguides (*105*), which had an index of 2.2 and losses for the TM_0 mode of 3 dB/cm at $\lambda = 0.63$ μm.

Ion-beam-sputtered films were prepared by bombarding a tantalum target with a mixture of argon and oxygen. For IBS films, indices of 2.18 at 530 nm (*44, 47*) and 2.03 (*99*) to 2.12 (*101*) have been reported. The extinction coefficient at 0.53 μm, for films prepared with adequate oxygen in the beam, was less than the 10^{-4} resolution of the fringe measurement technique (*117*). At 1.06 μm, the absorption of a half-wave layer was determined to be 78×10^{-6} (*101*).

(*2*) *Composition*. The composition of IBS films, as a function of the oxygen content in the sputtering beam, is an important parameter in determining the optical performance of the beam. Large shifts in the Ta XPS peaks between the metallic and fully oxidized states were used to determine the degree of oxidation. Approximately 44% O_2 was required in the beam to produce stoichiometric oxide films (*100, 117*). It was found that significant chemical reduction of the sample was incurred by the argon ion milling performed to clean the surface for the analysis, suggesting that a volume technique such as RBS would be helpful in the analysis of these films (*47, 117*).

(*3*) *Microstructure*. The films deposited by IBS were reported to be amorphous (*47, 101*), as determined by XRD and electron microscopy. Attempts to reduce the compressive stress by bombardment with a second source were not successful (*99*), but the films were reported to have good adhesion in single and multiple layers.

e. Aluminum Oxide (*Al_2O_3 Alumina*). Alumina forms either sapphire or another hexagonal crystal (corundum) with indices near 1.76, or a microcrystalline material with a similar index. It also has reduced and hydrated forms. In thin films, it is amorphous, with an index of 1.6 to 1.65. Additional oxygen is often used during deposition to ensure stoichiometricy.

(*1*) *Optical properties*. The use of oxygen ions during the evaporation of aluminum was used to produce oxide films in 1976 (*21*). Attempts to deposit Al_2O_3 with concurrent oxygen bombardment, however, resulted in an increase in the absorption of the films (*17*). Later work by Saxe (*112*) confirmed that reduction by the ion beam could result in increased absorption. A range of indices was observed, depending on the deposition conditions. Excess oxygen in the films was correlated with a decrease in the index. The current dependence of the index was investigated by McNally *et al.*, and they found a maximum of the index in the vicinity of 20 $\mu A/cm^2$ for both 500 and 1000 eV bombardment (*119*). IAD films tested as waveguides at $\lambda = 0.63$ μm were reported to have an index of 1.65 and losses of 1.5 dB/cm when deposited with 100 $\mu A/cm^2$, 1200 eV oxygen bombardment (*105*). These are comparable to the values of 0.5–1 dB/cm and $n =$ 1.65 reported for reactively sputtered waveguides (*120*). Indices between 1.65 and 1.7 at 0.63 μm have been reported for reactive ($Ar + O_2$) IBS films (*97*). These films had laser damage thresholds comparable to those observed in sputtered films.

(*2*) *Composition*. RBS analysis of IAD alumina films showed inclusion of both argon and oxygen. Excessive bombardment of the films led to the deposition of aluminum-deficient films, with the ratio of oxygen to aluminum rising as high as 3.14 : 2.0 (*112*). Argon concentration was seen to rise with the ion-to-molecule arrival ratio, γ; close to 3% inclusion was observed for $\gamma = 0.64$.

(*3*) *Microstructure*. Tests of the stability of alumina films have been made on exposure to air and fluorine environments (*112*), and their value as protective overcoats in acid baths has been assessed (*72*). In all of these cases, the improved performance of the IAD films over those deposited conventionally indicated a denser physical structure.

In oxygen bombardment of Al during deposition, it was found that the structure could be varied from an amorphous, fine-grained oxide to a cermet with small metal droplets in an oxide matrix (*21*).

f. Y_2O_3 *(Yttrium Oxide, Yttria)*. This material is of interest for short-wavelength laser systems, due to its relatively high index for a UV transmitting material. However, when evaporated, the indices of the films are highly inhomogeneous and unreproducible. It has been deposited by both ion-assisted deposition (*104, 121*), and ion-beam sputtering (*97, 104*). Films deposited with ion assistance were seen to have reduced inhomogeneity in the index, which was approximately 1.8–1.85 and constant through seven quarter-waves at 351 nm. In IBS-deposited yttria, overcoated Al mirrors were seen to have much less water-related IR absorption than those coated with conventionally *e*-beamed films. In both cases, a backpressure of oxygen was used to assure a stoichiometric deposition. Indices between 1.85 and 1.95 (at 0.63 μm) were reported by Varasi *et al.* (*97*) for Y_2O_3 ion-beam sputtered with an argon/oxygen beam. Laser damage thresholds for both IAD (*121*) and IBS (*97*) films were reported to be higher than for films deposited by other methods.

g. CeO_2 *(Cerium Dioxide)*. Cerium dioxide is another material for use as a high-index layer but also suffers from nonreproducible behavior and inhomogeneity of the index when evaporated conventionally. Recently (*122*) oxygen ion bombardment during deposition has been shown to increase the packing fraction and index, and to result in stable, reproducible films. For 1200 eV O_2^+, a maximum of the index was observed (greater than 2.4 at 0.55 μm) with an ion current density of about 200 $\mu A/cm_2$. A maximum in the index with fixed flux was observed for ion energies of 600 eV. The absorption in these films was greater than that of conventionally deposited films, and both the extinction coefficient and the scattering rose as the bombarding energy was increased (*122*). Measurements of the index and loss in waveguides yielded an index of 2.1 and 6 dB/cm at 0.63 μm (*105*).

h. i-C (Diamondlike Carbon). The use of ion beams for the production of extremely durable carbon films with good transmission characteristics was one of the most dramatic examples of coating improvements that are accessible with these techniques. Early results of Aisenberg (*16, 32*) and Spencer (*123*) using primary ion-beam sources encouraged the use of dual-ion-beam-sputtering techniques (*3*) for diamondlike carbon. Since carbon is not readily evaporated, limited use of IAD has been made for these films (*124*).

Using Ar as a carrier gas for the 40–60 eV C ions from a hollow cathode, Aisenberg and Chabot (*32*) reported the production of high-index (>2.0), mechanically durable, transparent coatings. When deposited on plastics, the abrasion resistance of the plastic was greatly enhanced (*16*). Transmission electron microscopic examination of similar C coatings by Spencer *et al.* (*123*) revealed a crystalline component, especially in films deposited on NaCl and KCl, and the lattice spacings observed corresponded to those for diamond. They confirmed the transparency in films as thick as 5000 Å.

The transmission of i-C coatings on Si in the infrared has been reported (*125*), and no excess absorption is seen from 2.5 to 40.0 μm. The coatings do not take up water, and serve as hermetic barriers (*125, 126*). The hardness is comparable to that of bulk diamond, and the adhesion to a variety of substrates is excellent (*127*).

Films deposited by dual-ion-beam sputtering, where a low-energy (500 eV) beam strikes a film being deposited by IBS, also show desirable mechanical and optical properties (*3, 128*). The structure is reported to be largely amorphous with the inclusion of small crystallites.

i. Other Dielectrics. Hafnium dioxide films deposited onto heated substrates have been reported to display an increase in the index of refraction at 0.35 μm from 2.02 to 2.11, using an oxygen ion flux of 12 $\mu A/cm^2$ at an energy of 500 eV (*119*). ZnS, ThF_4, AlF_3, and cryolite multilayers have been reported to have improved mechanical properties when deposited with 500 eV IAD (*23*). Low-energy bombardment of ZnS led to increased stress in thick layers (*83*).

2. Semiconductors

Lattice damage produced by the bombardment of the growing film in IAD has limited its use to the production of amorphous semiconductors, but IBS and ionized-cluster-beam deposition (ICB) have been shown to result in epitaxial growth at lower temperatures, due to the increased surface mobility of the arriving atoms. IAD has proven useful for doping of amorphous films during deposition, particularly for *a*-Si : H.

a. a-Si : H (Amorphous Hydrogenated Silicon). The compensation of the dangling bonds in amorphous Si films with hydrogen has made this material a practical choice for the production of low-cost, low-efficiency solar cells. Varying techniques for the introduction of the H have been investigated, among them IAD, IBS, and ICB.

The use of a mixture of hydrogen and argon ions to sputter a Si target results in doping of the amorphous film with significant amounts of H, leading to an increase in the effective energy gap. Early work by Weissmantel *et al.* (*20*) spurred further interest in the use of ion-beam doping techniques. Several workers have reported on the use of IBS, and the effects of varying the substrate temperature and bombardment energy (*91, 129, 130*). Singh *et al.* (*129*) reported variations in the gap from 1.4 to 1.9 eV could be achieved by variation of the temperature of the substrate and the ratio of hydrogen to argon in the beam. Low substrate temperatures and a 90% H beam were reported to result in almost complete saturation of the dangling Si bonds, with a final hydrogen concentration of 10% in the film (*128*). With a 1 keV beam and a substrate temperature of 220°C, the gap was varied from 1.8 to 1.95 eV as the hydrogen concentration increased from 65 to 80% (*130*). The gap was not strongly affected by the energy used for sputtering.

Martin *et al.* (*91*) made a comparison of reactively evaporated, IAD, and IBS films and found that the greatest increase in the gap could be achieved with IBS, using a substrate temperature of 300°C. They reported values of the index (in the visible) varying from 3 to 5 depending on the deposition method. IR absorption indicated the presence of some oxygen, and the films contained ≃15% hydrogen (*107*). Some densification of the films was indicated by a reduction in the vacuum-to-air shift of the transmission curves.

Reactive ionized-cluster-beam deposition *a*-Si : H resulted in thermally stable films. Si clusters evaporated in a backpressure of 10^{-4}–10^{-5} of H_2 gas had a mixture of monohydride and dihydride bonds; the monohydride could be favored by the use of higher acceleration voltages (*131*).

b. Epitaxial Si. Weissmantel (*128*) has reviewed the potential of ion beams (both IBS and ionized primary beams) for the production of single crystal Si films. While epitaxial growth can be achieved at temperatures of 800°C, there are a large number of electrically active defects which are introduced as a result of the energetic bombardment. Ionized-cluster-beam deposition in UHV has been reported to result in the production of epitaxial films at acceleration voltages of only 200 V. For high voltages, epitaxial growth was possible at lower temperatures in a diffusion-pumped vacuum (10^{-6} torr) without special cleaning procedures.

c. $In_xSn_yO_z$. Indium oxide and tin-doped indium oxide films are of interest as transparent conductors and are usually produced by reactive evaporation onto a heated substrate. Ebert (*17*) reported on the use of low-energy oxygen bombardment during deposition to fully oxidize In_2O_3 coatings. This reduced visible losses to less than 0.2% and eliminated the need for postdeposition thermal oxidation. Sheet resistances of 80 Ω/sq. were obtained with the addition of 30% Sn. The index of the doped films was 2.05. IAD of tin-doped indium oxide has also been reported by Dudonis *et al.* (*132*) and Martin *et al.* (*133*). The latter group used a 100 eV beam of oxygen during the evaporation of a mixture of In_2O_3 and Sn_2O_3. For room-temperature substrates, they reported an index of 2.13 and a sheet resistance of 800 Ω/sq. Increasing the substrate temperature to 400°C during deposition led to a decrease of the index to 2.0, and of the sheet resistance to 25 Ω/sq. Recently, reactive IBS of In has been used for the production of indium oxide and composites with small indium particles for use in optical data storage (*134*).

d. Other. Improved adhesion of other amorphous semiconductors, such as Ge–As–Se glasses (*82*) and Ge (*42, 135*) layers has been reported with the use of ion-beam techniques.

PbS and TiN deposited by ion-beam methods have been investigated because of their potential for use in solar energy conversion (*107, 136*). Values (over the visible) for the absorptance of 90% and emittance of 10% were reported for TiN, and about 70% and 8% for PbS. The values for PbS could be improved by the addition of an antireflection layer but were inferior to those obtained for evaporated films. Improvements in the deposition rate and mechanical properties were observed (*136*). Intermetallic compounds such as CdTe, PbTe, and Mn-doped CdTe have been deposited by ICB (*36*). Composition and crystallographic orientation were controlled by variation of the acceleration voltages.

3. Metals

Ion-beam techniques, particularly the ionized-cluster-beam method, have been applied extensively to the production of metal films. The use of ions in the processing can change the stress, adhesion, grain size, and optical response of metallic films.

a. Au, Ag. Films of the noble metals, particularly gold, are subject to delamination from the substrate, due to a limited chemical interaction between the metal and the oxide that comprises the substrate. The adhesion of gold and silver was reported to be improved (*23*) by the use of Ar and O bombardment prior to and during the deposition, with energies in the range of 0.5 to 1 keV. In particular, oxygen was found to be effective for gold if used during deposition. Martin *et al.* (*24*) observed similar effects and found that, after the formation of a bonding layer, the ion beam could be turned off. Early experiments by Marinov (*50*) on Ag indicated that the adatom mobility was enhanced, and that coalescence occurred at a smaller average thickness in IAD films. This effect has been employed in Au films by Smith (*137*) for the production of transparent IR reflecting layers. Extremely thin continuous layers have been reported by depositing Au with ion assistance, then milling away part of the coating (*138*).

rf-sputtered Cr films deposited with a dc bias during deposition were observed to have reduced columnar structure and smoother surfaces (*14*). Stress levels and reflectivity of Cr films were studied as a function of Ar and Xe bombardment, and it was observed that there was a bombardment flux threshold above which the reflectivity increased and the stress changes from tensile to compressive. Reduction of scatter has also been observed for Cu substrates overcoated with IAD Cu or Mo (*64, 65*).

Ion-beam sputtering has been applied to the production of a large number of metals (*96*). Changes in the crystal structure, mechanical, and optical properties result. The observed modifications correlate well with those observed for dc-sputtered metallic films.

Primary-beam deposition has also been used for the production of metal films (*33, 36, 115*). Direct deposition of a Pd beam (*36*) resulted in a higher density of nucleation sites, and large differences in the sticking coefficient were seen for insulating versus conductive substrates. Ionized-cluster-beam deposition results in continuous layers at smaller average thicknesses and some control over the surface morphology. Higher accelerating voltages for the beam led to smoother films (*115*). Measurements of oxygen concentration in Al films indicated a denser structure for films deposited at high voltages, and reflectance and crystallinity of Au and Cu films also increased.

VIII. Conclusion

Ion-beam processing has been applied to the production of a wide variety of materials for optical applications, with varying degrees of success. For metals and semiconductors, other techniques generally result in the production of comparable quality films, but for some of the dielectrics used in optical filters, ion-beam processing has proven highly advantageous. With separate control over the flux of the reactive ions and the amount of energy delivered to the growing film, both the stoichiometry and physical structure of the film can be tailored to meet design criteria.

Many of the undesirable characteristics of evaporated films can be eliminated using ion-assisted evaporation and ion-beam sputtering.

Perhaps the most significant change induced by these methods is the disruption of the natural columnar structure of refractory oxide films. The evolution of the fine-grained nucleation structure into these columns is though to be responsible for the inhomogeneity observed in some materials, and absorption of water into the voids between columns causes instabilities in the index, and even irreversible changes. Ion bombardment during deposition disrupts the growth of these columns, making a more homogeneous, stable layer. In addition, the ion-beam techniques often result in reduced crystallite size, yielding a more homogeneous structure on a local scale. For high-power applications, where any defect may result in film failure, production of amorphous material by these methods may prove desirable.

It is somewhat early in the development of ion-assisted evaporation to make generalizations, but patterns are emerging from the existing data and correlate with the models that we have for ion–surface interactions. For oxide films it is apparent that changes in the stoichiometry can be effected by low-energy bombardment, whereas changes in the crystal or columnar structure require more energy. This correlates well with the relative values of chemical activation versus sputtering and displacement energies. The disagreement in the magnitudes of the energies are largely due to the multiple collisions suffered by each incident ion.

The largest benefits of IAD are realized when the mobility of the adatoms is low, so that the additional energy supplied by the ion flux enhances surface migration appreciably. An additional factor may be the ability of the material to tolerate a wide range of bond lengths and angles between the constituent atoms. In those materials that can tolerate such variations, low mobility at the substrate results in an amorphous film, rather than a polycrystalline, columnar structure. For these materials (such as SiO_2 and Al_2O_3) a low-energy beam to assure stoichiometry may be sufficient. For materials which form small crystallites readily, then fail to bond to other nearby atoms, the columnar structure will be more pronounced. The addition of ion bombardment, which not only increases surface mobility, but also causes forward sputtering and subsurface rearrangement, should radically alter the structure of such a film. In addition, one would expect that a high substrate temperature combined with the ion bombardment would be desirable, to anneal out defects induced by the ions. This seems to be the case for materials such as TiO_2 and ZrO_2.

For several materials (TiO_2, ZrO_2, Al_2O_3), a maximum in the index of refraction has been observed with increasing flux at a fixed energy. This is presumably the result of a balance between sputtering and gas incorporation effects and the desired rearrangements near the surface. For each material, there should exist a set of optimum conditions, and careful investigations to map out the parameter space are required. Eventually, correlations of these observations with known chemical properties of the materials will shed more light on the details of the processes occurring in ion-assisted deposition.

References

1. P. J. Martin, *J. Mater. Sci.* **21,** 1 (1986).
2. T. Takagi, *Thin Solid Films* **92,** 1–17 (1982).
3. G. Gautherin and C. Weissmantel, *Thin Solid Films* **50,** 135 (1978).
4. J. M. E. Harper, J. J. Cuomo, R. J. Gambino, and H. R. Kaufman, *in* "Beam Modifications of Materials" (O. Auciello and R. Kelly, eds.), vol 1, Chapter 4. Elsevier, Amsterdam, 1984.
5. J. E. Greene and S. A. Barnett, *J. Vac. Sci. Technol.,* **21,** 285 (1982).
6. J. M. E. Harper, *in* "Thin Film Processes" (J. L. Vossen and W. Kern, eds.), Chapters 2-5. Academic Press, New York, 1978.
7. V. G. Bauer, *Am. Phys.* (*Leipzig*) [5] **19,** 434 (1934).
8. A. H. Pfund, *J. Opt. Soc. Am.* **24,** 99 (1934).
9. H. K. Pulker, "Coatings on Glass," p. 203. Elsevier, Amsterdam, 1984.
10. G. K. Wehner and G. S. Anderson, *in* "Handbook of Thin Film Technology" (L. V. Maissel and R. Glang, eds.), Chapter 3. McGraw-Hill, New York, 1970, L. I. Maissel, *ibid.,* Chapter 4.
11. G. S. Anderson, W. N. Mayer, and G. K. Wehner, *J. Appl. Phys.* **33,** 2991 (1962).
12. P. D. Davidse and L. I. Maissel, *J. Appl. Phys.* **37,** 574 (1966).
13. D. M. Mattox and G. J. Kominiak, *J. Electrochem. Soc.* **120,** 1535 (1984).
14. R. D. Bland, G. J. Kominiak, and D. M. Mattox, *J. Vac. Sci. Technol.* **11,** 671 (1984).
15. W. Heitmann, *Appl. Opt.* **10,** 2414 (1971).
16. S. Aisenberg and R. W. Chabot, *J. Vac. Sci. Technol.* **10,** 104 (1973).
17. J. Ebert, *Proc. SPIE—Int. Soc. Opt. Eng.* **325,** 29 (1982).
18. T. H. Allen, *Proc.—Int. Soc. Opt. Eng. SPIE* **325,** 93 (1982).
19. Chr. Weissmantel, *Thin Solid Films* **92,** 55-63 (1982).
20. Chr. Weissmantel, K. Bewilogua, D. Dietrich, H. J. Erler, H. J. Hinneberg, S. Klose, W. Nowick, and G. Reisse, *Thin Solid Films* **72,** 19 (1980).
21. J. Dudonis and L. Pranevicius, *Thin Solid Films* **36,** 117 (1976).
22. Kaufman-type sources are available from Ion Tech, Fort Collins, Colorado and Commonwealth Scientific, Virginia.
23. W. C. Hermann, Jr. and J. R. McNeil, *Proc. SPIE—Int. Soc. Opt. Eng.* **325,** 101 (1982).
24. P. J. Martin, W. G. Sainty, and R. P. Netterfield, *Appl. Opt.* **23,** 2668 (1984).
25. D. W. Hoffman and M. R. Gaerttner, *J. Vac. Sci. Technol.* **17,** 524 (1980).
26. M von Ardenne, *in* "Tabellen der Elektronenphysik, Lonenphysik und Ubermicroskopie," p. 554. Dtsch. Verlag Wiss, Berlin, 1956.
27. H. R. Kaufman and P. D. Reader, *Am. Rocket Soc.* [pap.] **1374-60** (1960); H. R. Kaufman, *Adv. Electron. Electron Phys.* **36,** 265 (1974).
28. Cold cathode sources are available from, for example, Denton Vacuum Inc., Cherry Hill, New Jersey.

29. J. Amano, P. Bryce, and R. P. W. Lawson, *J. Vac. Sci. Technol.* **13,** 591 (1976).
30. S. Aisenberg and R. Chabot, *J. Appl. Phys.* **42,** 2953 (1971).
31. I. Yamada, H. Takaska, H. Inokana, H. Usui, S. C. Cheng, and T. Takagi, *Thin Solid Films* **92,** 137 (1982).
32. S. Aisenberg and R. Chabot, *J. Appl. Phys.* **42,** 2953 (1971).
33. J. Amano, *Thin Solid Films* **92,** 115 (1982).
34. I. Yamada and T. Takagi, *Thin Solid Films* **80,** 105 (1981).
35. I. Yamada, H. Takoaka, H. Inokawa, H. Usui, S. C. Cheng, and T. Takagi, *Thin Solid Films* **92,** 137 (1982).
36. T. Takagi, *J. Vac. Sci. Technol., A* [2] **2,** 382 (1984).
37. E. G. Spencer and P. H. Schmidt, *J. Vac. Sci. Technol.* **8,** 552 (1971).
38. O. Auciello and R. Kelly, eds., "Beam Modifications of Materials," Vol. 1. Elsevier, Amsterdam, 1984.
39. K. L. Chopra, *in* "Thin Film Processes," 164, (R. E. Krieger, ed.), p. 164. McGraw-Hill, New York 1979 (originally published 1969).
40. E. Taglauer, W. Heiland, and U. Beitat, *Surf. Sci.* **89,** 710 (1979).
41. R. V. Stuart and G. K. Aldiner, *J. Appl. Phys.* **33,** 2345 (1962).
42. E. H. Hirsch and I. K. Varga, *Thin Solid Films* **69,** 99 (1980).
43. H. M. Naguib and Roger Kelly, *J. Nucl. Mater.* **35,** 293 (1970).
44. U. Gibson, unpublished.
45. R. Kelly, *Surf. Sci.* **100,** 857 (1980).
46. H. Demiryont and J. R. Sites, *J. Vac. Sci. Technol., A* [2] **2,** 1457 (1984).
47. S. M. Rossnagel and J. R. Sites, *J. Vac. Sci. Technol., A* [2] **2,** 376 (1984).
48. K. L. Chopra, *in* "Thin Film Processes" (R. E. Krieger, ed.), p. 181. McGraw-Hill, New York, 1979 (originally published 1969).
49. N. Kanekama, N. Taniguchi, K. Watanabe, M. Kondo, and T. Matsumoto, *Sci. Pap. Inst. Phys. Chem. Res. Jpn.* **67,** 25 (1973).
50. M. Marinov, *Thin Solid Films* **46,** 267 (1977).
51. R. Kelly, *in* "Beam Modifications of Materials" (O. Auciello and R. Kelly, eds.), Vol. 1, p. 81. Elsevier, Amsterdam, 1984.
52. K. H. Müller, *J. Vac. Sci. Technol.* **A4**(2), 184 (1986).
53. K. H. Müller, *J. Appl. Phys.* **59,** 2803 (1986).
54. P. J. Martin, H. A. Macleod, R. P. Netterfield, C. G. Pacey, and W. G. Sainty, *Appl. Opt.* **22,** 178 (1983).
55. H. M. Liddel, "Computer Aided Techniques for the Design of Multilayer Filters," Chapter 6, p. 119. Adam Hilger, Ltd., Bristol, 1981.
56. A. Hjortsberg, *Thin Solid Films* **69,** L15 (1980).
57. L. Vriens and W. Rippens, *Appl. Opt.* **22,** 4105 (1983).
58. W. E. Case, *Appl. Opt.* **22,** 1832 (1983).
59. J. Mouchard, G. Lagier, and P. Pointu, *Appl. Opt.* **24,** 808 (in French) (1985).
60. J. C. Manifacier, J. Gasiot, and J. P. Fillard, *J. Phys. E* **9,** 1002 (1976).
61. M. Hacskaylo, *J. Opt. Soc. Am.* **54,** 198 (1964).
62. T. Tamir, *Top. Appl. Phys.* [2] **7,** 84ff (1979).
63. K. H. Guenther, P. G. Wierer, and J. M. Bennett, *Appl. Opt.* **23,** 3820 (1984).
64. J. R. McNeil, G. A. Al-Jamaily, K. C. Jungling, and A. C. Barron, *Appl. Opt.* **24,** 486 (1985).
65. G. A. Al-Jamaily, J. J. McNally, J. R. McNeil, and W. C. Hermann, *J. Vac. Sci. Technol., A* [2] **3,** 651 (1985).
66. M. J. Messerly, H. A. Macleod, J. A. Leavitt, and J. D. Targove, *Opt. News* **11,** p. 122 (1985).

67. G. E. McGuire, *in* "Deposition Technologies for Films and Coatings" (R. F. Bunshah, ed.), Chapter 14. Noyes, Park Ridge, New Jersey, 1982.
68. L. S. Hsu, C. Y. She, and G. J. Exharos, *Appl. Opt.* **23,** 3049 (1984).
69. L. S. Hsu, R. Rujkorakarn, J. R. Sites, and C. Y. She, *J. Appl. Phys.* **59,** 3475 (1986).
70. K. H. Guenther, *Appl. Opt.* **23,** 3497 (1984).
71. S. G. Saxe, M. J. Messerly, B. Bovard, L. DeSandre, F. J. Van Milligan, and H. A. Macleod, *Appl. Opt.* **23,** 3633 (1984).
72. W. G. Sainty, R. P. Netterfield, and P. J. Martin, *Appl. Opt.* **23,** 1116 (1984).
73. M. L. Stein, S. Aisenberg, and J. Stevens, Post-deadline paper presented at CLEO, Washington, D.C. (1981).
74. E. Spiller, I. Haller, R. Feder, J. E. E. Baglin, and W. N. Hammer, *Appl. Opt.* **19,** 3022 (1980).
75. D. S. Campbell, *in* "Handbook of Thin Film Technology" (L. I. Maissel and R. Gland, eds.), Chapter 12. McGraw-Hill, New York, 1970.
76. T. Lichtenstein, Master's Thesis, Institute of Optics, University of Rochester, Rochester, New York, (1980).
76a S. D. Jacobs, A. L. Hrycin, and C. Baldwin, *Lab. Laser Energetics Review* **22,** 85 (1985).
77. R. Brown, *in* "Handbook of Thin Film Technology" (L. I. Maissel and R. Glang, eds.), Chapter 6, p. 41. McGraw-Hill, New York, 1970.
78. V. Bennett and J. Smith, *Proc. Soc. Photo-Opt. Instrum. Eng.* **126,** 76 (1977).
79. A. R. Bayly and P. D. Townsend, *Opt. Laser Technol.* **8,** 117 (1970).
80. J. R. McNeil and W. C. Hermann, *J. Vac. Sci. Technol.* **20,** 324 (1982).
81. L. F. Johnson and K. A. Ingersoll, *Appl. Opt.* **22,** 1165 (1985).
82. W. C. Hermann, Jr. and J. R. McNeil, *Proc. 11th Annu. Symp. Opt. Mater. High Power Lasers, Boulder, Colorado, 1980.*
83. D. Kuchibhatla, Optical Sciences Center, University of Arizona, Tucson (personal communication) (1985).
84. R. Kelly *in* "Beam Modifications of Materials" (O. Auciello and R. Kelley, eds.), Vol. 1, pp. 31, 35. Elsevier, Amsterdam, 1984.
85. M. C. Wittels and F. A. Sherrill, *Phys. Rev.* **93,** 1117 (1954).
86. K. L. Chopra, K. S. Harshavardhan, S. Rajagopalan, and L. K. Malhotra, *Appl. Phys. Lett.* **40,** 428 (1982).
87. B. Zhang, J. M. Delavaux, and W. S. C. Chang, *Appl. Opt.* **23**(9), 777 (1984).
88. E. Garmire, *Top. Appl. Phys.* [2] **7,** 263, 273 (1982).
89. J. Bartella *et al., Appl. Opt.* **24,** 2625 (1985).
90. W. C. Hermann, Jr. and J. R. McNeil, *Appl. Opt.* **20,** 1899 (1981).
91. P. J. Martin, R. P. Netterfield, W. G. Sainty, and D. R. Mackenzie, *Thin Solid Films* **100,** 141 (1983).
92. J. R. McNeil, G. A. Al-Jumaily, K. C. Jungling, and A. C. Barron, *Appl. Opt.* **24,** 486 (1985).
93. H. Küster and J. Ebert, *Thin Solid Films* **70,** 43 (1980).
94. J. R. McNeil, Alan C. Barron, S. R. Wilson, and W. C. Hermann, *Appl. Opt.* **23,** 552 (1984).
95. T. H. Allen, *Proc. Int. Ion Eng. Congr., 1983,* p. 1305, (1983).
96. K. L. Chopra and M. R. Randlett, *Rev. Sci. Instrum.* **38,** 1147 (1967).
97. M. Varasi, C. Misiano, and L. Lasaponara, *Thin Solid Films* **17,** 163 (1984).
98. T. H. Allen, *Optics News* **11,** 131 (1985).
99. J. R. Sites, P. Gilstrap, and R. Rujkorakarn, *Opt. Eng.* **22,** 447 (1983).
100. H. Demiryont and J. R. Sites, *Proc. 16th Annu. Symp. Opt. Mater. High Power Lasers, Boulder, Colorado, 1984* (1984).

101. J. R. Sites, H. Demiryont, and D. B. Kerwin, *J. Vac. Sci. Technol., A* [2] **3,** 656 (1985).

102. G. A. Al Jumaily, S. R. Wilson, A. C. Barrow, J. R. McNeil, and B. L. Doyle, *Nucl. Instrum. Methods Phys. Res. Sect. B* **7,** 906 (1985).

103. K. Yamashini, Y. Minowa, and A. Shuhara, *Proc. Int. Ion Eng. Congr., 1983,* p. 1203 (1983).

104. B. E. Cole, T. J. Moravec, R. G. Ahonen, and L. B. Ehlert, *J. Vac. Sci. Technol., A* [2] **2,** 372 (1984).

105. L. N. Binh, R. P. Netterfield, and P. J. Martin, *Appl. Surf. Sci.* **22/23,** 656 (1985).

106. P. J. Martin, R. P. Netterfield, and W. G. Sainty, *J. Appl. Phys.* **55,** 235 (1984).

107. P. J. Martin, R. P. Netterfield, W. G. Sainty, and C. G. Pacey, *J. Vac. Sci. Technol., A* [2] **2,** 341 (1984).

108. P. J. Martin, R. P. Netterfield, W. G. Sainty, G. J. Clark, W. A. Lanford, and S. H. Sie, *Appl. Phys. Lett.* **43,** 711 (1983).

109. U. J. Gibson and C. M. Kennemore, III, *Thin Solid Films* **124,** 27 (1985).

110. C. M. Kennemore, III and U. J. Gibson, *Appl. Opt.* **23,** 3608 (1984).

111. C. M. Kennemore, III and U. J. Gibson, *Opt. News* **10,** 78 (1984).

112. S. G. Saxe, Ph.D. Dissertation, University of Arizona, Tucson (1985).

113. P. J. Martin and R. P. Netterfield, *Appl. Opt.* **24,** 1732 (1985).

114. U. J. Gibson and C. M. Kennemore, III, *Proc. SPIE* **678,** 130 (1986).

115. T. Takagi and I. Yamada, *Appl. Opt.* **24,** 879 (1985).

116. S. D. Jacobs, A. L. Hrycin, K. A. Cerqua, C. M. Kennemore, III, and U. J. Gibson, *Thin Solid Films* **144,** 69 (1986).

117. H. Demiryont, J. R. Sites, and K. Geib, *Appl. Opt.* **24,** 490 (1985).

118. J. R. McNeil, private communication (1985).

119. J. J. McNally, G. A. Al-Jumaily, and J. R. McNeil, *Opt. News,* p. 131 (1985).

120. G. Este and D. Westwood, *J. Vac. Sci. Technol., A* [2] **2,** 1238 (1984).

121. D. J. Smith, Masters report, Optical Sciences Center, University of Arizona, Tucson (1985).

122. R. P. Netterfield, W. G. Sainty, P. J. Martin, and S. H. Sie, *Appl. Opt.* **24,** 2267 (1985).

123. E. G. Spencer, P. H. Schmidt, D. C. Joy, and F. J. Sansalone, *Appl. Phys. Lett.* **29,** 118 (1976).

124. S. Fujimori and K. Nagai, *Jpn. J. Appl. Phys.* **20,** L194 (1981).

125. S. Aisenberg, *J. Vac. Sci. Technol., A* [2] **2,** 369 (1984).

126. S. Aisenberg and M. Stein, *Proc. SPIE—Int. Soc. Opt. Eng.* **299,** 64 (1981).

127. J. Koskinen, J. P. Hirvonen, and A. Anttila, *Appl. Phys. Lett.* **47,** 941 (1985).

128. D. J. Smith and C. Weissmantel, *Proc. Int. Ion Eng. Congr, Kyoto, 1983,* p. 1257 (1983).

129. J. Singh, R. C. Budhani, and K. L. Chopra, *J. Appl. Phys.* **56,** 1097 (1984).

130. Y. Suzuki, S. Ogawa, K. Taguchi, and H. Matsuda, *Proc. Int. Ion Eng. Congr., Kyoto, 1983,* p. 845 (1983).

131. I. Yamada, M. Horie, and T. Takagi, *Proc. Int. Ion Eng. Congr., Kyoto, 1983,* p. 1197 (1983).

132. J. Dudonis, A. Jotatis, O. Meilus, A. Meskauskas, and L. Pranevicius, *Thin Solid Films* **58,** 106 (1979).

133. P. J. Martin, R. P. Netterfield, and D. R. Mackenzie, *Thin Solid Films* **137,** 204 (1986).

134. A. F. Hebard, G. E. Blonder, and S. Y. Suh, *Appl. Phys. Lett.* **44,** 1023 (1984).

135. E. H. Hirsch and I. K. Varga, *Thin Solid Films* **52,** 445 (1978).

136. P. J. Martin, R. P. Netterfield, and W. G. Sainty, *Thin Solid Films* **87,** 203 (1982).

137. G. B. Smith, *Appl. Phys. Lett.* **46,** 716 (1985).

138. P. J. Martin, R. P. Netterfield, W. G. Sainty, and C. Pacey, *Pap., Am. Vac. Soc. 32nd Nat. Symp., 1985,* Pap. No. JS2THM9 (1986).

Laser-Induced Etching

Carol I. H. Ashby

Sandia National Laboratories
P.O. Box 5800
Albuquerque, New Mexico 87185

I. Introduction

The utilization of thin films in an increasing number of technologies imposes certain potential restrictions on the type of fabrication processes which can be used. Many applications, especially those in microelectronics, require the selective removal of small areas of a thin layer of one material from the underlying substrate material to produce a particular pattern on the surface. This must often be done with a minimum of energy deposition in the surrounding material to avoid impurity diffusion, thermal degradation, or damage. A great variety of wet processes using liquid etchants and dry processes using gaseous etchants have been developed to etch thin films. Most, however, will etch the entire film surface unless additional masking steps are employed to spatially restrict access of the reactant to produce the desired pattern. The spatial localization provided by a laser may be the solution to the problem of etching patterns into thin-film layers in a highly controlled manner while reducing the number of process steps.

Many of the current dry processes used to pattern thin-film materials employ ions to remove material directly, e.g., ion-beam milling or sputtering, or to enhance the rate of chemical etching reactions, e.g., reactive ion etching (RIE) and ion-beam-assisted etching (IBAE). Laser-based processes possess some general characteristics which provide advantages over these other etching techniques for certain applications. Unlike ion-based processes, laser processes seldom produce atomic displacement damage or defects which can require postprocessing annealing for electronic materials. Lasers can be used to etch materials with a wide variety of etchant sources, including liquids and both low- and high-pressure gases. Discrimination between layers of different chemical composition is possible. Material removal rates can be very high (>1 μm/min) or quite low (0.001 μm/min).

A wide variety of laser-based processes will be discussed in this article, with special emphasis on those for materials with current applications in microelectronics. Due to the vast body of work in this area, this review is not exhaustive, but great care has been taken to make it broadly representative of the possibilities inherent in laser-induced etching. The general characteristics, advantages, and disadvantages of the different methods of laser-induced etching are the primary emphasis of this review. However, a survey of much of the work reported to date on specific materials is also provided. Most of the work reviewed here deals specifically with laser etching and has been reported since 1976, since most of the effort in this field has been quite recent. However, it is obvious that many photochemical reactions involving solids may be directly applicable to laser-induced etching as well.

II. Mechanisms of Laser-Induced Etching

The mechanisms involved in laser-induced etching can be divided into four general categories: ablation, thermal, gas-phase or liquid-phase photochemical, and solid-phase photochemical. Much of the work through 1981 involving examples of thermal and gas- or liquid-phase photochemical processes has been reviewed by Chuang (*1*) from the standpoint of the fundamental mechanisms involved. Specific examples of all four process types will be presented for a variety of materials in Section IV.

1. Ablation

Ablation occurs when high laser-power densities, i.e., greater than 10^6 W/cm^2, are employed to rapidly etch a solid without deliberate introduction of a chemically reactive species. This process normally is performed under vacuum or in air and is applicable to any type of material. Ablation has been employed to etch polymers, insulators, semiconductors, and metals. However, it shows little if any selectivity between materials of different chemical composition. Except when applied to polymers, ablation-based etching processes tend to leave a significant amount of redeposited material, or debris, on the edge of the etched region.

Significant effort has been expended in studying the ablation of polymers. Photoablation of polymers is based on breaking a very large number of bonds within a very short time; this results in a volume expansion and ejection of material. After a critical number density of broken bonds is reached, a large increase in volume produces explosive ejection of material at speeds in excess of 1000 m/sec (*2, 3*). This ejected material is vibrationally, rotationally, and translationally excited (*4, 5*). If the laser-power density is sufficiently high, a plasma is formed at the surface (*6*). The critical factor for producing ablation is the laser power density, i.e., the rate of deposition of energy, not just the total energy deposited. A threshold power density for significant material removal is generally observed (*6, 7*). It is common for authors to report this as a threshold fluence, with a constant laser pulse length being an unstated assumption. A lower total energy density is required to etch a given depth if the power density is higher, as illustrated in Fig. 1 (*8*). The process is relatively independent of substrate temperature (*4*).

At high power densities, the rate is independent of whether the ambient is vacuum, inert gas, or air (*7, 9, 10*), but the volatile products are altered (*7, 11–13*). At lower power densities, rates are higher in the presence of O_2 (*4, 9*). With a scanned cw beam, no threshold energy is observed and O_2 is required for total film removal (*14*). The oxygen probably reacts with some of the radicals produced by bond photolysis and prevents bond reformation. Stress transients up to 10^7 Pa occur in the substrate due to rapid material ejection (*15*). Stress measurements indicate some ablation occurs even at "subthreshold" fluences (*15*).

There has been some controversy over the relative importance of pho-

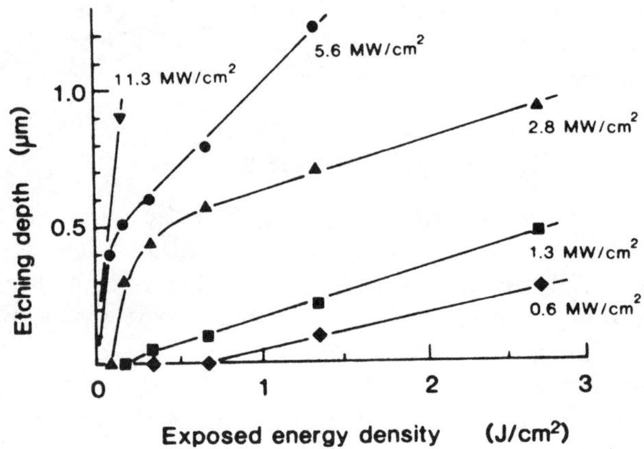

FIG. 1. UV photoetching characteristics of PMMA (containing a small amount of benzoin) for a KrF excimer laser (248 nm). [From Kawamura *et al.* (*8*).]

tochemical and photothermal bond breaking when short-wavelength uv light (193 nm) is used. The general concensus has been that ablation with light of 248 nm or longer wavelength is primarily due to thermal bond breaking (*3, 7, 16–18*). Intense local heating is proposed to rupture polymer bonds. Localized material temperatures in excess of 1000 K have been reported (*17*). This photopyrolysis mechanism is consistent with the observed wavelength dependence of threshold fluence (*7*). Since 193 nm light is strongly absorbed by many polymers, the possibility for photolytic as well pyrolytic bond breaking exists. Although a primarily photolytic mechanism for ablation with 193 nm light was proposed for some time, recent studies where the thermal energy loading of a polymer was measured for fluences from below threshold to well above threshold have shown that thermal processes play an important role at all wavelengths (*17, 18*). Calorimetric studies show that 80–90% of the deposited energy above threshold is carried away by the ejected material (*18*). Pulses of acoustic energy are measured in the solid due to momentum transfer from the ejected material (*15, 18*). A mechanistic analysis in terms of polymer degradation kinetics has been reported for 193 nm ablation (*19*).

For materials other than polymers, ablation is generally attributed to a laser-induced phase change where the substrate is volatilized by the rapid energy deposition. For metals, there is some evidence that melting and not direct vaporization is the principal phase change involved in the ablation process (*20*).

2. Thermal Processes

In thermal processes, the laser provides heat to the solid in a highly localized way to increase the rate of a thermally activated chemical reaction. This approach is applicable to any type of material for which there are steps in etching reactions which are strongly temperature dependent; consequently, there are great opportunities for developing a wide variety of laser-induced thermal etching processes. Enhancement may result from the temperature dependence of either product formation or product desorption. Material selectivity is limited only by the chemical selectivity of the particular etching reaction employed. Much lower laser power densities may be employed for this type of process since the temperature of the solid may need to rise only a few tens of degrees to produce substantially higher etching rates in the irradiated area. However, the fastest reactions involve laser-induced melting of the solid. The rise in the temperature of the solid can be a consequence of electronic or vibrational excitation followed by thermalization during relaxation to the ground state. The temperature rise resulting from laser irradiation has been modeled theoretically (*21–26*) and has been measured experimentally by a number of techniques (*27*).

Some thermal processes have been identified for metals such as Al, where temperature gradients serve simply to crack the passivating oxide layer and do not increase the rate of a thermally activated chemical process. Rapid etching occurs because the cracks permit access of reactants such as Cl to the underlying reactive Al surface (*28, 29*).

3. Gas-Phase and Liquid-Phase Photochemical Processes

In gas- or liquid-phase photochemical processes, the laser is used to generate the reactive species responsible for etching the solid. In general, a reactant precursor molecule is chosen which strongly absorbs the laser photons and then decomposes to produce the actual reactant, which has a much higher reactivity toward the solid than the original molecule. The precursor may or may not be adsorbed on the surface when the light is absorbed. This approach is applicable to any type of material, and selectivity is limited only by the chemical selectivity of the particular chemical reaction. Most of the work in this area has employed gas-phase reactants (*30*), although this is primarily due to the desire to use dry etching processes rather than to any fundamental limitation of liquid-phase reactants themselves. The laser beam can be either parallel or perpendicular to the solid surface. When it is perpendicular, there is often a thermal component to the role of the laser in the reaction, and the laser induced etching proceeds at a substantially higher rate.

4. Solid-Phase Photochemical Processes

Solid-phase photochemical processes involve direct electronic excitation of the solid surface by the incident light. This type of process has been restricted to semiconductor materials. The excitation produces additional free electrons and holes in the semiconductor. These photogenerated carriers are responsible for the increase in reactivity under laser irradiation. There are two general categories of solid-phase photochemical etching processes. The first type is photoelectrolytic or photoelectrochemical (PEC) etching. These processes require ohmic contacts to the semiconductor and an external power supply. Etching is controlled by the applied current density while photons are used to create a plentiful supply of both electrons and holes in the surface region. The solid to be etched is immersed in an electrolyte solution, but the chemical properties of the electrolyte are not usually of fundamental importance to a PEC process. The second type does not require an external current source. In contrast to PEC etching, the chemical properties of the solution or gas mixture are very important in determining whether etching occurs and at what rate. With liquid etchants, the illuminated and unilluminated regions of the semiconductor surface serve as an anode and a cathode for a galvanic reaction. Material selectivity is determined by the chemical selectivity of the particular etching reaction.

A process based on electronic excitation of the solid can display a high level of selectivity between materials with very similar chemical reactivity by using the electronic properties of the materials to control the generation or subsequent behavior of the photogenerated carriers which are responsible for the etching. Since photogenerated carriers participate directly in the etching process, the effect of various recombination processes such as surface recombination and bulk recombination play a vital role in determining the efficiency of a solid-phase photochemical process.

III. Methods of Laser-Induced Etching

There are four major process characteristics which are of both fundamental and practical concern in etching patterns into solid surfaces: resolution, etching rate, etching selectivity, and substrate damage. The resolution of the laser-induced etching process can be of great importance for practical applications, especially for microelectronic materials, which currently require micrometer-scale feature definition and for which submicrometer resolution is ultimately desired. For some manufacturing applications, the rate of the etching process may be a crucial factor in determining total throughput on a production line. The ability to selectively etch some materials while not etching others is of obvious importance in the fabrication of layered structures, such as those found in many electronic devices. Finally, the damage to the surrounding area produced by the etching technique must be minimal. Different laser-based methods for producing a pattern on a solid surface and their general characteristics in terms of resolution, rate, selectivity, and damage are discussed in this section.

1. Resolution

The minimum feature size which can be etched by a laser-induced process may be determined either by the optical properties of the apparatus or by characteristics of the etching process. Depending on the process, the actual resolution which can be achieved may be higher or lower than the optical limit. Purely optical effects of wavelength, spatial and temporal coherence, and different methods of image formation, will be considered in Section III,1,a. Process effects on the ultimate resolution will be considered in Section III,1,b.

a. Optical Limitations on Resolution

(*1*) *Wavelength.* Although the ultimate optical resolution of a laser-etching process is determined by the wavelength and coherence of the laser and the optical imaging elements employed, the most fundamental limiting factor is the wavelength. The shortest wavelengths permit the highest optical resolution. However, other process considerations, such as selectivity or minimizing damage, may make it more advantageous to use longer wavelengths (see Sections III,3 and III,4). A longer wavelength may also be desirable if absorption of shorter-wavelength light in the surrounding medium degrades the optical characteristics of the system by "thermal blooming" (*31*) or bubble formation. Commercially available lasers which have been used extensively for laser etching and their principal wavelengths are listed in Table I. The wavelengths in parentheses are achieved using frequency-doubling techniques. These lasers in turn can be used to pump dye lasers to produce other wavelengths that are longer than the pump-laser wavelength. The use of Raman shifting of the wavelength of excimer lasers provides a way to achieve wavelengths shorter or longer than the original excimer output (*32*). This ability to

TABLE I

PRINCIPAL WAVELENGTHS OF COMMONLY AVAILABLE LASERS

Mode	Laser	Wavelength[a]
CW	He–Ne	633
	He–Cd	442
	Kr^+	799, 753, 676, 647, 568, 531, 521, 483, 476, 356, 351
	Ar^+	529, 514.5, 488, 477, 458, 364, 351, (257)
	Nd–YAG	1064, (532)
Pulsed	F_2	157
	ArF	193
	KrF	248
	XeCl	308
	XeF	351
	N_2	337
	Ruby	694
	Nd–YAG	1064, (532), (266)
	CO_2	10.6 μm, 9.10–11.1 μm (tunable)

[a] Units of nm unless specified otherwise.

select the wavelength for the laser-etching process is responsible for the high degree of selectivity among similar materials exhibited by some processes, as will be discussed in Section III,3.

(2) *Coherence*. Some laser sources, such as HeNe, Ar^+, Kr^+, and Nd–YAG, have extremely coherent output. It is possible to operate these lasers in a single mode with a low degree of angular divergence of the output beam. Consequently, these lasers can be used to produce useful patterns, such as gratings, on the surface of a substrate by periodically varying the intensity of light at the surface using the interference of two beams produced by splitting the original laser output beam. This method of pattern formation is discussed in III,1,a,(1). However, it is this very coherent output which make these lasers undesirable for other imaging techniques and for some processes which require very low light intensity. In addition, the well-known phenomenon of laser speckle is generally undesirable. Coherence reduction techniques (*33*), such as passing the beam through a hollow-tube waveguide (*34*), have been employed to eliminate some of these problems. In contrast, excimer lasers, such as ArF, KrF, and XeCl, exhibit low spatial and temporal coherence. Their larger spectral bandwidth (100–200 cm^{-1}) and high output divergence make them good light sources for patterning techniques based on image projection and contact or proximity masking (Sections III,1,a,(1)).

(*3*) *Pattern-Formation Techniques*. There are four general approaches to pattern formation which have received considerable attention to date: image projection, contact or proximity masking, direct writing, and interference techniques. Each of these will be dealt with in turn in this section.

Image Projection. This technique and the technique of contact masking (Section III,1,a,(3)) are the techniques normally employed in conventional photolithographic patterning processes. In the image-projection technique, a mask is situated remotely from the substrate surface, and an image of the mask is demagnified by projection optics to give the desired pattern at the surface. A common approach employs Schwartzchild reflecting optics to produce the demagnified image (*31*). The use of reflecting optics rather than refracting optics is especially desirable when short-uv-wavelength sources are used, since this avoids the problem of light absorption in the optical elements. A pattern can also be formed in a projection mode by placing the surface to be etched in the Fourier plane (*35*). The diffraction-limited feature size for image projection is given by

$$w = \lambda/2(\mathrm{NA}) \tag{1}$$

where λ is the wavelength in the medium in contact with the substrate and NA is the numerical aperture [NA = optical element diameter/(2 × focal length)]. A large numerical aperture is desirable to get transmission of the higher spatial frequencies (*31*). It is important to have a highly planar substrate surface when working near the diffraction limit because the depth of field, Z, given by

$$Z = \lambda/(\mathrm{NA})^2 \tag{2}$$

is very shallow under these conditions (*36*). Submicrometer-sized features have been produced regularly using image-projection techniques. For example, gratings with 0.2 μm spacings have been produced on Si using 193 nm light and COF_2 (*37*). When image-projection techniques have been used to etch organic materials at power densities in excess of 2 MW/cm^2, a loss of resolution has been observed (*31*).

Contact or Proximity Masking. With contact or proximity masking, the mask which defines the pattern is in direct contact with the substrate or separated by only a very small distance. Proximity masking uses a stencil mask, which is not directly deposited on the substrate. This approach may be used for patterning photoresists in conventional, i.e., nonlaser, photolithographic processes. Patterned photoresist films on a substrate surface are an example of contact masking for subsequent processing steps. Virtually all indirect, i.e., two-step, patterning processes use this technique for subsequent pattern transfer to the substrate. Separate focusing optics for the laser light are not required.

The pattern may be formed by exposing a photoresist and subsequently developing the image, typically with a wet etching solution, or by direct, i.e., one-step, photochemical reaction. The resolution of this approach is given by

$$w^2 = \lambda z/2, \tag{3}$$

where z is the mask–surface separation (*36*).

Diffraction effects are a prominent characteristic of this technique when small feature sizes are involved. An example of the complications due to diffraction is provided by the PEC etching of InP using a metal proximity mask with dimensions where Fresnel diffraction effects are important (*38*). Light-guiding effects decrease the effective resolution as holes are etched deeper; trenching and undercutting occur, as shown in Fig. 2 (*38*). Trenches form because of the intensity maxima resulting from Fresnel diffraction. The movement of the Fresnel peaks away from the wall as the hole deepens causes tapering and trenching. This effect depends on the etch depth and the diffraction geometry but not on the light intensity. Undercutting occurs because of the nonzero intensity of light in the shadow of the mask. This problem is worsened by higher light intensity. Trenching increases the severity of this problem by scattering light

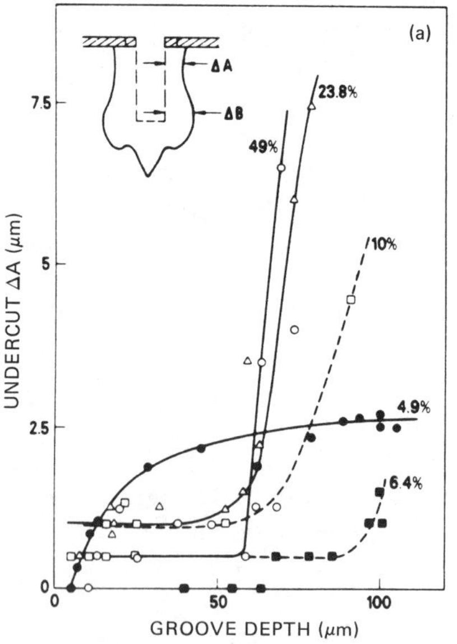

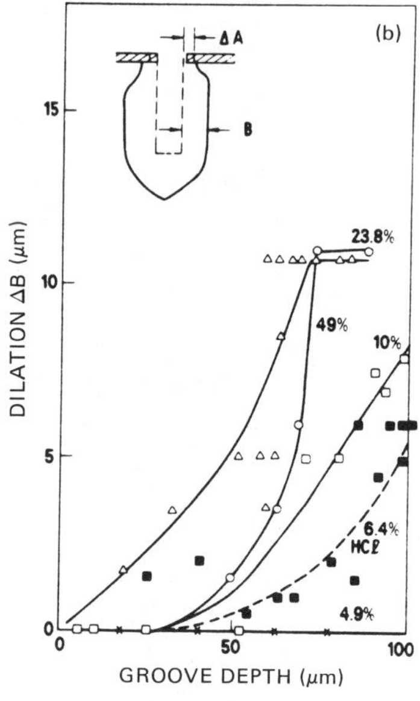

FIG. 2. Broadening of groove due to (a) mask undercut, and (b) wall erosion, as a function of the percentage of light transmitted, T. [From Cheng and Kohl (*38*).]

onto the sidewalls. At sufficiently high intensities, the rate of the PEC process at the hole bottom reaches a saturation value. Further increases in intensity have little effect on the etching rate for the hole bottom but continue to increase the rate of sidewall erosion. Thus, higher-intensity light may be detrimental to the resolution of the process.

High-resolution etching has been achieved with contact or proximity masking. Exposure of a methyl methacrylate–methacrylic acid copolymer with 157 nm light followed by wet development has produced 0.13 μm lines (*39*). Single-step etching of 0.3 μm grooves in Mo has been achieved using interference fringes from a single-narrow-slit proximity mask to photodissociate NF_3 with 193 nm light (*40*).

Direct Writing. Pattern formation by direct writing does not employ a mask. Rather, the output beam of the laser is directly focused on the surface of the substrate. The beam may be held stationary, producing a hole in the substrate, or scanned, producing a line. There are two general categories of direct-write processes. In a one-step process, material is removed while illuminated by the laser. In the two-step process, the laser is used to expose a resist which is then removed by a subsequent development step, as in conventional photolithography. The arrangement of focusing optics determines whether the spot at the surface is diffraction limited or not (*41*); advanced autofocusing techniques minimize this problem. A diffraction-limited spot size can be achieved, but this requires that the final optical element be located very close to the substrate surface. In addition, the very shallow depth of field requires a very flat surface (*41*). Longer working distances may be used; this results in a larger minimum spot size, although the depth of field is increased. Typical optical arrangements for these two approaches are illustrated in Fig. 3 (*41*). For a Gaussian-profile beam, the diffraction-limited spot size at $1/e^2$ of the maximum intensity is given by

$$w = \lambda/n(\mathrm{NA}) \tag{4}$$

where λ is the vacuum wavelength and n is the index of refraction of the medium in contact with the substrate surface (*41*). Immersion of the last optical element in a solution with a high index of refraction produces a smaller spot size. Comparison of the ultimate resolution of image projection and one-step and two-step direct writing requires knowledge of particular process characteristics (*42*) and will be discussed in Section III,1,b,(3).

Interference Techniques. The earliest patterning techniques used in laser-induced etching employed the modulation of light intensity which occurs when a beam is split into two beams that are subsequently allowed to interfere with each other at the substrate surface. This approach is

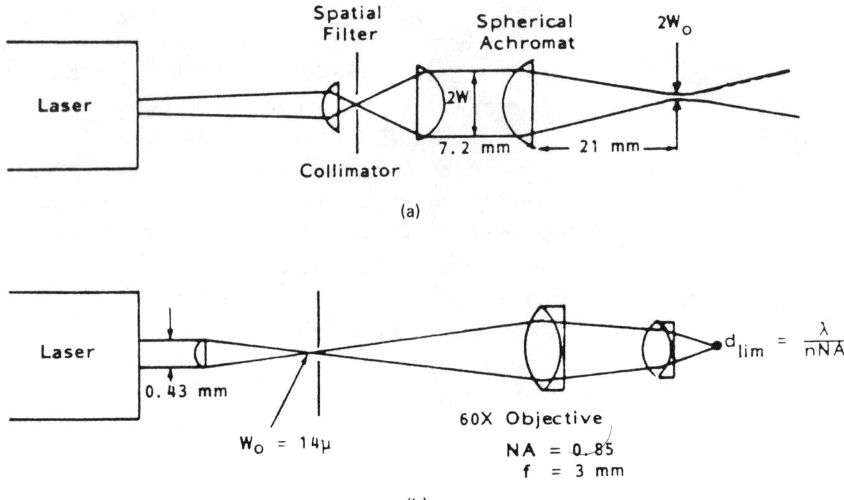

FIG. 3. Focusing optics for laser spot scanning. (a) Long working distance, large spot size; (b) short working distance, small spot size. [From Kavassalis *et al.* (*41*).]

limited to the formation of highly periodic structures such as gratings. The relief gratings produced by these interference techniques are classed as thin-film holograms, so the technique is also called holographic etching. The interference of the two beams produces a sinusoidal intensity variation at the surface. The spatial frequency of this light pattern is determined by the angle between the two beams. The spacing between intensity maxima is given by

$$a = \lambda/2n \sin \theta \tag{5}$$

where λ is the vacuum wavelength, n is the index of refraction of the etching medium, and 2θ is the angle between the two beams. The grating period is changed by varying θ.

There are two important components to the resolution of an interference technique. The first is the spatial frequency, and the second is the amplitude modulation, which is the difference in the height of the maxima and minima of the grating. It is desirable to maximize this difference, and grating depths which are 20% of the separation between lines have been reported (*43–45*). In general, the amplitude modulation decreases with prolonged etching (*46*). This degradation is faster when higher-intensity light is used. The quality of a grating is generally better if it is etched at lower intensities; in fact, a long-time etch at higher intensities can completely destroy a grating (*46*). Carrier diffusion is an important factor in limiting both the spatial frequency and the amplitude modulation, as will be discussed in Section III,1,b,(2).

The quality of a grating can be monitored during etching by monitoring the intensity of the first-order diffracted beam (*47*). This first increases with etch time, then decreases (*45, 46, 48*). To get the most efficient grating for the diffraction of light in air, the etching time that produces the maximum intensity in the etching solution should be adjusted by the relative refractive indices of the two media (*48*).

Gratings with nonsinusoidal profiles can be made by minor alterations in the optical system. Blazed gratings can be made by tilting the substrate relative to the bisector of the angle between the interfering beams (*45*). Undercut profiles are obtained by tilting the substrate out of the plane of the interferometer (*49*). Gratings with submicrometer spacings tend to have a cusplike profile (*45*).

The maximum number of lines in a grating, N, is determined either by the bandwidth or the source divergence:

$$N = 2\lambda/\Delta\lambda \tag{6}$$

$$= 2/(\Delta\theta)^2 \tag{7}$$

where $\Delta\lambda$ is the bandwidth or wavelength spread and $\Delta\theta$ is the divergence angle (*50*).

A special technique known as spatial frequency doubling is an alternative to regular interference techniques (*50, 51*). The divergence and bandwidth are less critical in determining N, but the necessity of suppressing the zeroth order places more rigid constraints on the optical configuration. Using 193 nm light, 99.5 nm grating spacings have been achieved (*51*).

b. Process Limitations on Resolution. Because laser-induced etching relies on some physical or chemical process to remove the substrate material, the particular characteristics of a given process may decrease or increase the ultimate resolution from that predicted by optical considerations alone. Some of the most common factors that can alter the resolution are reactant diffusion, carrier diffusion, and temperature-dependent processes.

(*1*) *Reactant Diffusion.* When a laser-etching process involves the photogeneration of reactive chemical species, the distance through which these highly reactive species diffuse before undergoing reaction or being otherwise deactivated can affect the resolution of the laser-etching process. This effect can be especially important with gas-phase reactants.

From the gas kinetic theory for a single gas, the mean free path r of a molecule is given by

$$r = 1/[2^{1/2}\pi n d^2(1 + C/T)] \tag{8}$$

where n is the number density, d is the molecular diameter, T is the temperature (K), and C is Sutherland's constant, which is a measure of the attraction between molecules and typically has a value between 50 and 600 (*52*). The diffusion-limited linewidth is given by the sum of the laser-beam diameter and two times the mean free path (*27*). The mean free path is reduced by increasing either the pressure or the temperature. With pressures on the order of an atmosphere, collisional deactivation and recombination have been demonstrated to contain the reactive species within the irradiated zone and produce diffraction-limited features (*40, 53*). One can provide such a high pressure either with the gas which is the precursor of the photogenerated reactant or with a buffer gas. However, if the gas is too efficient in promoting deactivation, there may be an induction time before etching proceeds at a useful rate (*54*). With liquid-phase reactants, the inherently higher density of molecules ensures that the diffusion length for reactive species will be short. For example, the mean free path for halogen atoms produced from I_2 or Br_2 in aqueous solution is less than 30 nm (*55*). Consequently, optical considerations determine the resolution for liquid-phase reactions.

(2) *Carrier diffusion.* The diffusion of photogenerated carriers can degrade the ultimate resolution of etching processes that are dependent on the number density of carriers at the surface. This is true for solid-phase photochemical etching of semiconductors with either liquid or gaseous reactants, since the etching rates for these processes tend to be linearly dependent on the minority carrier concentration at the surface. The effect of carrier diffusion has been most extensively studied in interference-based etching of gratings. Limits on the spatial frequency and modulation amplitude in n-type materials are largely determined by diffusion and recombination of holes (*43–45, 56, 57*). Lateral diffusion of holes along the surface has been shown to explain the observed decrease in modulation amplitude and, consequently, of diffraction efficiency at high spatial frequencies. Factors which decrease the hole lifetime at the surface improve the resolution. This is clearly seen in the comparison of the spatial frequency response of gratings in n-InP and undoped InP in Fig. 4 (*56*). The higher electron concentration in n-type material produces faster electron–hole recombination, thereby reducing lateral hole diffusion. Diffusion of holes can still be a controlling factor for small feature dimensions even in heavily doped n-type materials. The decrease in amplitude modu-

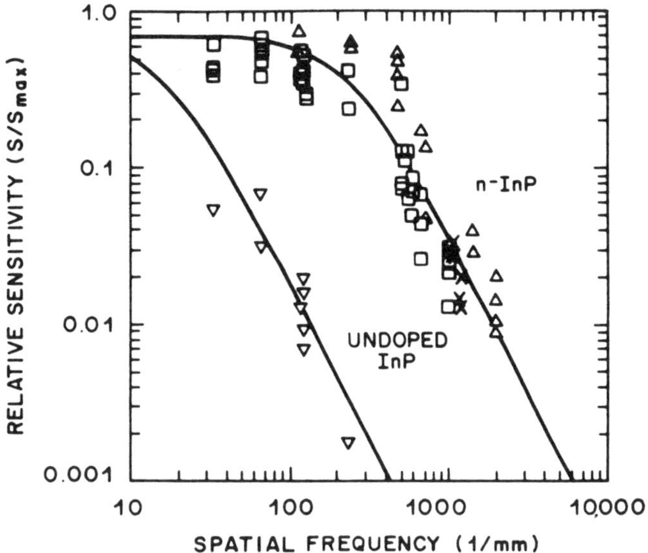

FIG. 4. Spatial frequency response of gratings in n-InP and undoped InP for incident beam wavelengths of 441.6 (△), 488 (X), and 632.8 nm (□, ▽). [From Lum *et al.* (*56*).]

lation, as demonstrated by the diffraction efficiency in Fig. 5 (*56*), is pronounced as one goes from a grating spacing of 8.5 to 1.5 μm.

Higher reaction rates can also decrease the time during which holes can diffuse laterally (*43, 57*). Increasing the concentration of reactants by a factor of 10 markedly increases the reaction rate and improves the modulation amplitude for n-InP by a factor of 100 at a spatial frequency of 500/mm, as illustrated in Fig. 6 (*57*). A rapid wet etch of n^+-GaAs has produced gratings with an optical interference resolution of 100 nm using 257 nm light (*58*). At high light intensities, space-charge effects may become important and may degrade resolution by increasing lateral hole flow (*30*).

(2) *Thermal effects on resolution.* There are two broad categories of laser-induced etching processes that use a laser to heat the substrate. The first includes chemical reactions that are thermally activated, and the second involves thermal degradation-based photoablation, as discussed in Section II,1. Thermal processes tend to be nonreciprocal; i.e., they depend not just on the total energy deposited in the material but also on the rate at which the energy is deposited. Consequently, it is possible to produce features which are either larger or smaller than the original beam diameter, depending on the total energy deposited, the energy deposition rate, and the thermal-diffusion properties of the substrate.

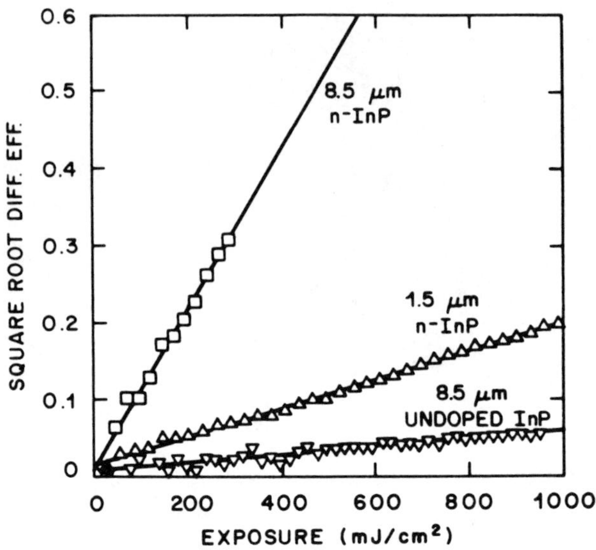

FIG. 5. Measurements of the holographic sensitivity from the linear portion of the square root of the diffraction efficiency versus exposure curves for gratings in *n*-InP and undoped InP. [From Lum *et al.* (*56*).]

In general, narrower features are produced by pulsing or rapidly scanning the beam, while wider features are produced if the beam is stationary or scanned slowly. This is possible because thermally activated chemical reactions exhibit an exponential dependence on temperature, and the temperature profile of the substrate under laser irradiation depends on the rate of energy deposition. Extensive work in modeling the substrate temperature profiles under a variety of irradiation conditions have been reported. The temperature profiles produced with a stationary cw Gaussian-profile beam have been calculated (*22–23*). Since the thermal conductivity normally decreases and the absorption coefficient normally increases with increasing temperature, calculations show that the laser-power density required to produce a significant temperature rise may be substantially lower than expected (*23*). Modeling of the temperature rise with pulsed lasers has shown the same effects (*25, 59, 60*). In addition, the pulse shape, pulse duration, and energy density affect the temperature rise and the thermal gradient profile (*59, 60*). Similar effects are observed when a cw beam is rapidly scanned across the substrate (*24, 61*). The change in temperature profile along the beam path with increasing scanning speed is illustrated in Fig. 7 (*61*).

Narrow features are favored when there is a steep thermal gradient in

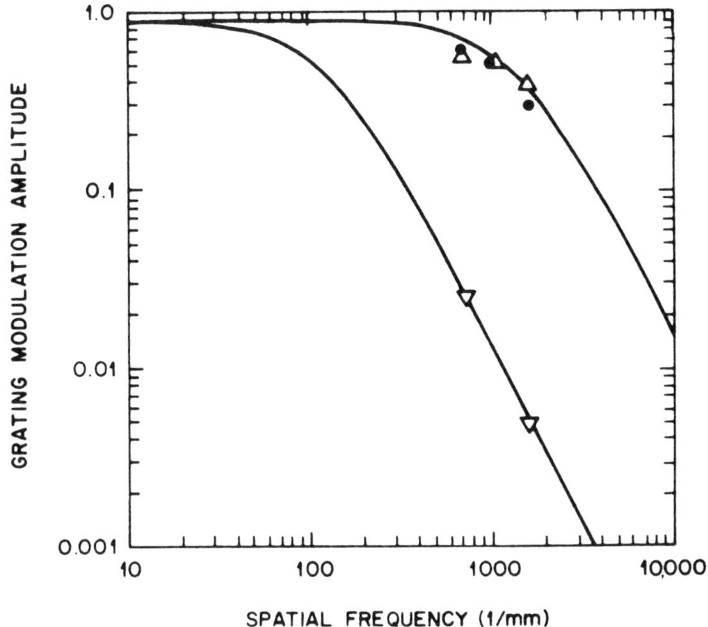

FIG. 6. Spatial frequency response of gratings etched in *n*-InP using $H_2SO_4/H_2O_2/H_2O$ (△, ratio 1 : 1 : 10; ▽, ratio 1 : 1 : 100) and H_2SO_4/H_2O (●, ratio 1 : 10) electrolytes. [From Lum *et al.* (*57*).]

the substrate. This is produced either by rapid scanning (*62, 63*) or by pulsing a stationary beam (*63*). For example, a laser has been used to heat the surface of GaAs to produce a reaction with CCl_4 (*64*). A 0.6 μm wide line was etched using a 1.2 μm beam and a scan speed of 30–60 μm/sec. Feature width can be made narrower or wider than the beam diameter by varying the scanning speed. When GaAs was etched with a 24 μm beam, linewidths varied from 7 μm at a scan speed of 12 μm/sec to 70 μm at a scan speed of 3 μm/sec (*64*). An alternative to increasing scan speed is reducing laser power (*66*). With a stationary cw Gaussian-shaped beam, one gets conical hole shapes due to thermal diffusion. An example of such a conical hole profile is provided by the VIA etched through a 250 μm wafer of Si with a 5 μm beam; the hole was 40 μm in diameter at the entrance of the hole and 15 μm at the exit (*62*). When reaction rates are high, calculations predict that an exothermic reaction may release as much heat as the laser heat input and markedly alter the temperature profile (*67*).

One-step direct writing using a thermally activated chemical process with a fairly large activation energy, and therefore a large contrast param-

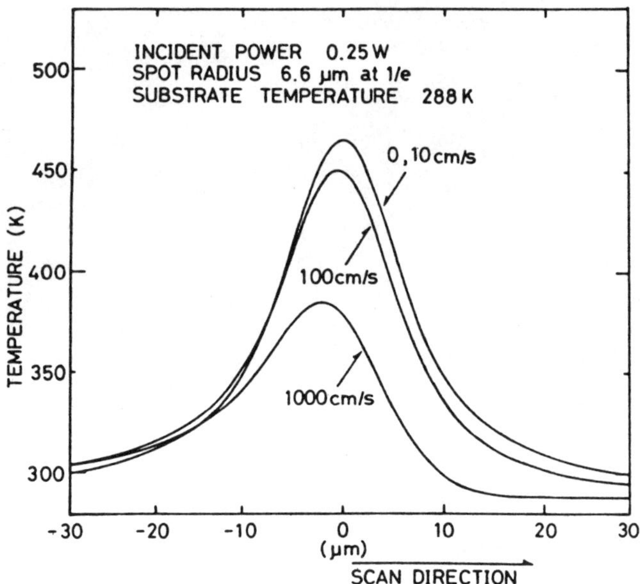

FIG. 7. Temperature profile along a beam scanning direction for scanning speeds of 0, 10, 100, and 1000 cm/s. [From Takai *et al.* (*61*).]

eter, can produce a minimum line spacing which is one-fourth that possible using image projection (*42*). A two-step direct-writing process has a resolution between that of one-step direct writing and image projection, depending on its nonreciprocity parameter (*42*).

Some additional thermal effects that can result from high thermal gradients in the etching medium rather than at the surface may degrade the resolution. These include changes in the index of refraction of the medium with temperature and turbulence. In addition, bubble formation from localized boiling can be a problem with liquid etchants.

2. ETCHING RATES

a. Temporal Characteristics of Light Source. Either cw or pulsed lasers can be used to produce high etching rates; the best choice depends on the particulars of the materials, the process, and the necessary resolution. Pulsed sources are desirable for processes requiring high peak-power densities ($>10^6$ W/cm^2), such as ablation and many thermally activated chemical reactions. Pulsed sources can produce both high material-removal rates and high resolution. Although there is a very high etching rate during the pulse, the low duty cycle can result in a much slower time-integrated rate than a comparable cw process. For example, for the same average power density, the etchings of GaAs by HNO_3 using 257 nm cw light is five times faster than when using 30 Hz 248 nm pulsed light (*68*). Very fast rates have been reported with cw light. Rates on the order of 10 μm/sec have been reported for the reaction of Si while molten using 10^7 W/cm^2 514 nm light in the presence of Cl_2 (*62, 69*). When high-intensity cw sources are used, however, the resolution is often lower than with a comparable pulsed source due to thermal diffusion. For photochemical processes, which are linearly dependent on the number of photons, cw sources are preferable to pulsed sources since higher rates can be achieved with a minimum of substrate heating.

For thermal reactions with a scanned cw beam, rates are not necessarily linearly proportional to the beam dwell time. For the thermal reaction of GaAs with CCl_4, a beam dwell time of 2 sec produces an etch rate of 9 μm/min, while a dwell time of 8 sec produces a rate of 300 μm/min (*65*). This is due in part to the nonlinear temperature dependence of the thermal reaction rate. In addition, the substrate is not in thermal equilibrium at short dwell times, so the time-dependent rise in temperature is important (*24*).

Sensitivity to power instabilities in the laser output is lower at higher scanning speeds (*24*). For reactions involving a molten surface, it is easier to control the melt depth by varying the scanning speed than by varying the laser-power density (*24*). Hot spots in a pulsed beam can produce uneven etch profiles; this can be minimized by using a beam homogenizer to produce a more uniform intensity profile (*33, 34*).

Thermal gradients in the reactant medium near the surface are sometimes desirable to increase the reaction rate, since they can produce a stirring effect to provide a continuous supply of fresh reactant. Thermal gradients are more pronounced when using a pulsed or chopped beam. When using a chopped cw beam, this effect manifests itself as a maximum in etching rate as a function of increasing chopping period (*70*).

b. Geometry.

(1) Small spot versus semi-infinite plane. If the maximum rate is limited by the diffusion of reactants to the substrate surface, the diffusion-limited reaction rate for a small spot may be as much as 10^4 times faster than the rate for a semi-infinite surface (*30*). Thermal and stress gradients contribute to this increase in rate. As the spot size decreases, a much higher flux of reactant molecules is possible before the reaction becomes diffusion limited, as illustrated in Fig. 8 (*30*). When the substrate is located in a

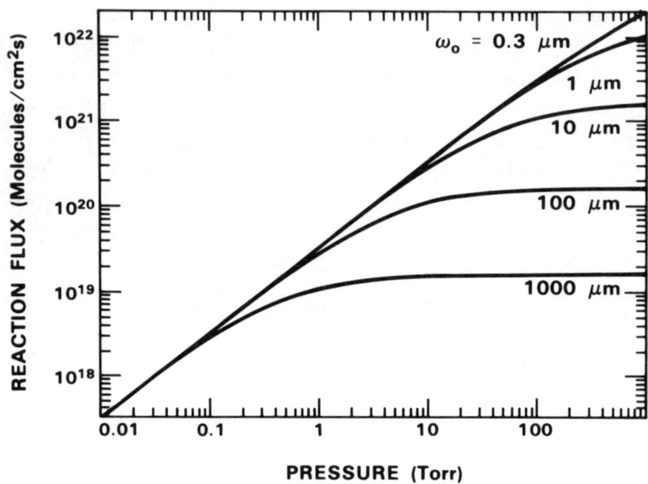

FIG. 8. Steady-state values for the molecular reaction flux as a function of pressure for various values of the spot radius, ω_0, with $T = 1000°C$ and diffusivity and pressures typical of thermally activated laser etching. [From Ehrlich and Tsao (*30*).]

flowing stream of reactant at one atmosphere, a 3 μm diameter beam would be expected to produce a reaction rate which is 10^4 times faster than the rate for a large-area surface (*30*).

(*2*) *Aspect ratio*. When the ratio of the depth of a hole to its width, i.e., the aspect ratio, exceeds one, a change in the reaction rate often occurs. Depending on the particulars, there may be either a decrease or an increase in rate, although the former is more common. A common cause of the decrease in rate with increasing aspect ratio is a limitation on the transport of reactants to the surface or products from the surface. This is seen in the PEC etching of InP, where the limited availability of reactant in deep, narrow grooves results in a current-saturation effect (*38*). Additional effects sometimes occur, however. For instance, the etch depth per pulse for the polymer PMMA decreases at high aspect ratio. This is probably due to the absorption of uv light by the ablated monomer which has not cleared the hole during the laser pulse (*71*). Under appropriate conditions, wave-guiding effects can prevent the decrease in rate from occurring or even increase the rate, as will be discussed in the next section.

c. Wave-Guiding Effects. Wave-guiding or light-guiding effects occur when the feature walls reflect the incoming beam deeper into the substrate. The effect is well illustrated by the etching of a 1 μm VIA through a 100 μm thick GaAs wafer using wet etching with 257 nm light at 100 mW/

cm^2 (*72*). A different study demonstrated the etching of 1.5–6 μm straight-wall vias through GaAs using a 3 μm beam of 257 nm light at intensities ranging from 10 mW/cm^2 to 1 kW/cm^2 (*73*). This was accomplished using both focused beams and projected patterns. Using a Gaussian-shaped TEM_{00} beam, the hole profile was initially Gaussian. As the hole deepened, non-Gaussian wall profiles developed due to the angular dependence of the Fresnel reflectivity and surface tilt. Vertical-walled tubes formed. The rate was 2–3 times faster than the initial rate until the tube was approximately 100 μm deep. Then the rate slowed due to reactant or product transport limitations. For small-diameter deep holes, the etch rate using 257 nm light remained higher at greater depths than the rate with 514.5 nm light. This occurs because the attenuation of intensity is less at the shorter wavelength, since the intensity attenuation is proportional to λ^2 divided by the hole radius cubed (*73*). The near-metallic reflection of uv light by GaAs facilitates wave-guiding effects. Wave guiding also plays an important role in etching deep features in CdTe by photosublimation (*74*).

In contrast, for photoablation etching of polymers such as PMMA and polycarbonate, reflection from the walls can drop the intensity below the threshold fluence required for photoablation. This results in a higher rate at the beam center and can lead to nonuniform hole profiles, as shown in Fig. 9 (*71*). The problem can be prevented in materials which have a fairly low threshold fluence, such as PMMA, by using higher laser fluences.

d. Wavelength Dependence. The rate of laser-induced etching can depend strongly on the wavelength of light used. In this section, wavelength-dependent etching that is not solid-phase photochemical etching of semiconductors is discussed. Since the wavelength dependence of semiconductor etching is related to carrier-generation effects, this will be discussed in Section III,2,e.

Gas-phase photochemical etching can be performed with ir, uv, or visible light. In general, IR-driven processes require absorption of many photons within a very short period of time and, consequently, require high-intensity light sources to produce useful etching rates. This is clearly illustrated by the etching of Si or Ta by SF_6 (*75, 76*). Three or more photons are required to produce reactive vibrationally excited SF_6 (*75*); alternatively, 30 or more photons may be required to photodissociate SF_6 to produce SF_4 and F atoms (*76*). Such multiphoton processes exhibit a nonlinear dependence on laser intensity, and high-intensity light sources are required to produce etching. Visible- and UV-driven processes may be either multiphoton or single photon in nature. In all cases, however, gas-phase photochemical processes require that the precursor molecules have

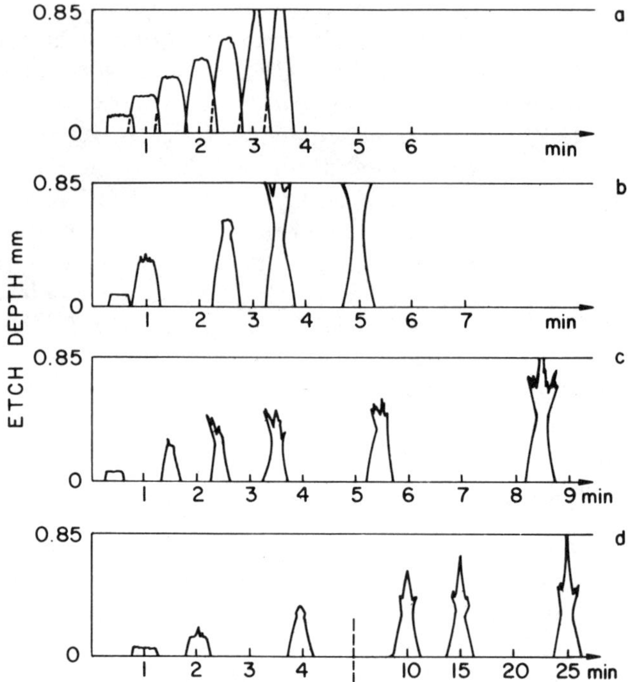

FIG. 9. Experimental profiles of holes drilled in PMMA with 193 nm laser radiation. Aspect ratio at end 4.3; focal length of lens 15 cm; fluences: (a) 475, (b) 350, (c) 250, and (d) 150 mJ/cm^2; number of pulses per minute is constant (3 Hz). [From Braren and Srinivasan (*71*).]

a significant absorption cross section for photons of the laser wavelength if etching is to occur.

For the UV ablation of polymers, the threshold laser fluence for ablation is lower for shorter wavelengths, as illustrated in Fig. 10 (*7*). The etched depth per pulse D is given by

$$D = (1/\alpha)\ \ln(\phi_{\text{inc}}/\phi_{\text{th}}) \tag{9}$$

where α is the absorption coefficient at the laser wavelength, ϕ_{inc} is the incident laser fluence, and ϕ_{th} is the threshold fluence for etching at the laser wavelength (*7, 10, 16*).

Thermal process control has been demonstrated by laser heating of a layer of a material with a high absorption coefficient for the laser light and which has been grown on a transparent substrate (*77, 78*). While $Gd_3Ga_5O_{12}$ (GGG) is transparent at wavelengths greater than 490 nm, the Fe–garnets used in magnetic films often are highly absorbing at these

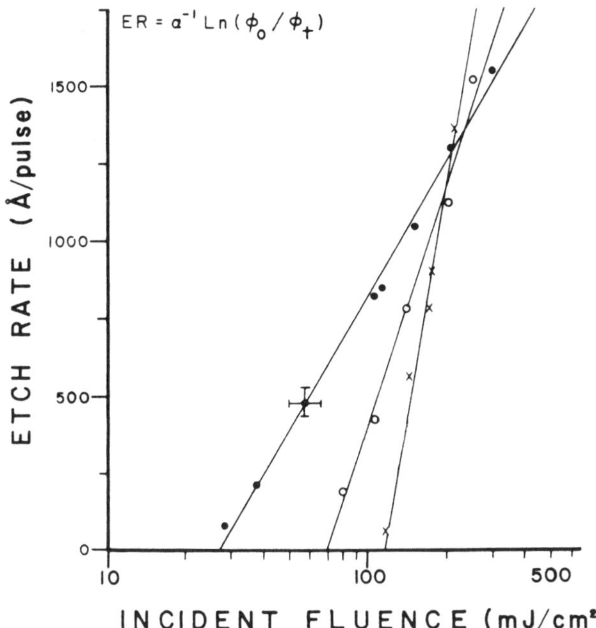

FIG. 10. Plots of experimentally measured single-pulse etch depth versus incident fluence. Etching was performed in air. (●), KrF, 248 nm; (○), XeCl, 308 nm; (X), XeF, 351 nm. [From Brannon *et al.* (*7*).]

wavelengths. This difference in absorption has been used with $(YBi)_3(Fe\text{-}Ga)_5O_{12}$ on GGG to heat the Fe–garnet layer with a multiline Ar^+ laser to produce etching with H_3PO_4. Etching proceeds down to the GGG substrate and stops, leaving only a few Fe–garnet islands (*77*). A more complicated application has employed the same principle to level Fe–garnet films of different compositions to a desired thickness (*78*). In this case, etching occurs until the increase in temperature from absorption in the thin layer is insufficient to produce significant etch rates.

e. Carrier Generation and Control. The necessity of significant concentrations of photogenerated carriers at the surface of the substrate provides excellent opportunities for control of solid-phase photochemical etching of semiconductors. Some of the approaches which produce a high degree of selectivity among chemically similar materials will be discussed in Section III,3.

The wavelength dependence of the etching of semiconductors is often related to the difference in the spatial distribution of photogenerated carriers in the materials. For example, the laser-enhanced plasma etch rates of $10^{16}/cm^3$ *p*-Si are the same with 350, 514, and 647 nm light (*35, 79*). In contrast, for $10^{20}/cm^3$ *p*- or *n*-Si, the etch rates decrease in the order 350 > 514 > 647 nm (*35*), as illustrated in Figs. 11 and 12. This effect can be explained by the shallower depletion depth for heavily doped materials and the decrease in photon absorption depth at lower wavelengths (*35, 79*). Since photogenerated carriers are swept toward the surface by the field present in the depletion region, the shallower depletion depth of a heavily doped material may not permit collection of photons generated deeper in the material, thereby lowering the etching rate. Similar results have been reported with both *n*-GaAs(100) (*58*) and Cr-doped semi-insulating GaAs(100) (*80*) using acid/peroxide wet etching solutions. For GaAs, this effect was moderated by increasing the laser intensity. While the rate with 257 nm light was 10 times greater than the rate with 514 nm light at intensities of 10 W/cm², the rates were nearly equal for intensities in excess of 10 kW/cm² (*72*). A role for hot, nonthermalized holes was proposed (*72*).

Recombination processes, which reduce the concentration of carriers available to participate in etching reactions, can markedly affect rates. The higher rates obtained with *p*- and *n*-Si compared to p^+-Si, as shown in Fig. 13, can be attributed to the faster bulk recombination in the more heavily doped material (*81*). An increase in surface recombination, as shown by a drop in the photoluminescence intensity, also correlates with a decrease in reaction rate (*82*). Ion-damaged regions etch more slowly

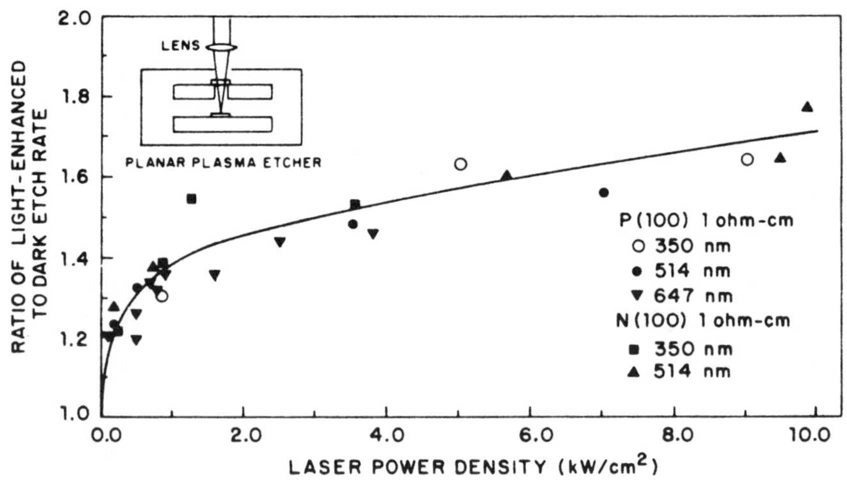

FIG. 11. Normalized photoinduced etch rate enhancement versus laser power density for $10^{16}/cm^3$ Si, etched in 120 mtorr CF_4/O_2. Laser wavelengths are indicated in the figure. [From Reksten *et al.* (*35*).]

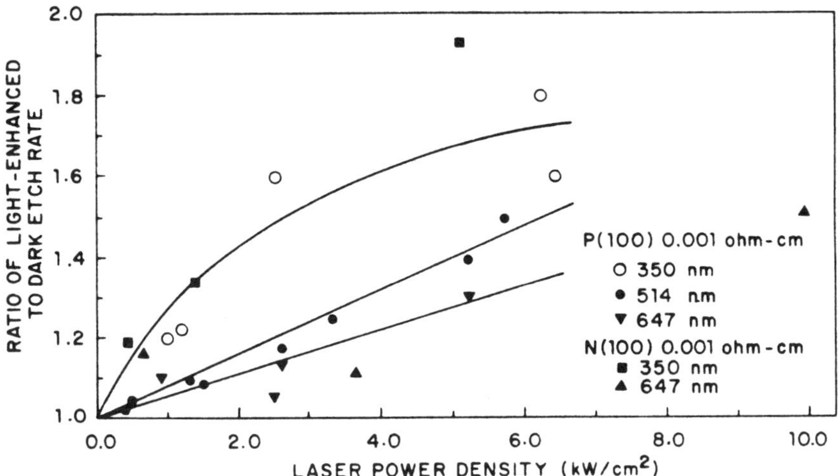

FIG. 12. Normalized photoinduced etch rate enhancement versus laser power density for $10^{20}/cm^3$ Si, etched in 120 mtorr CF_4/O_2. Laser wavelengths are indicated in the figure. [From Reksten *et al.* (*35*).]

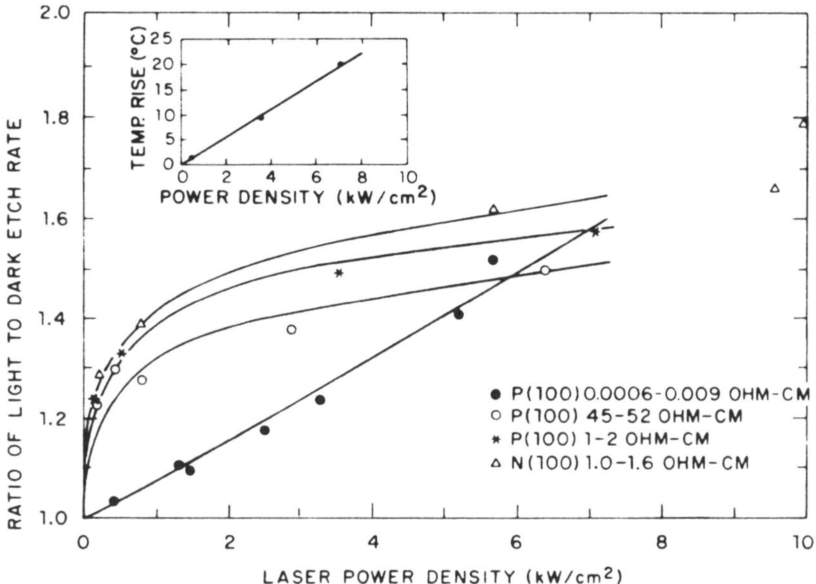

FIG. 13. Laser-induced etch enhancement versus laser-power density. The inset shows the calculated laser-induced temperature rise versus laser-power density for a 95 μm diameter focal-spot size. [From Holber *et al.* (*81*).]

under illumination than undamaged regions due to the increased carrier recombination at damage sites (*83, 84*). The effect of ion-beam damage on the PEC etching of III–V materials has been studied for GaAs, InGaAs, InGaAsP, and InP (*84*). It was shown that the change in etch rate was linear in ion dose from a threshold of approximately 5×10^9 ions/cm^2 to a dose which completely suppresses PEC etching ($3\text{–}7 \times 10^{10}$ ions/cm^2, depending on the material). Damage introduced by mechanical polishing of a semiconductor surface retards PEC etching of semiconductors (*41*).

For PEC reactions, etch rates are proportional to the minority carrier concentration at the surface. For *n*-type materials, these are holes. The rate is controlled by both light intensity and an applied bias voltage. The light controls generation of minority carriers, and the bias voltage controls minority carrier flow to the surface. This approach is exemplified by the PEC etching of *n*-Si (*85*) and of *n*-InP (*57*). For InP, no wavelength dependence was seen, even though there was a factor of four difference in the absorption depth (*56*). This may be attributed to the control of the depth from which carriers are collected by the applied bias. It is desirable to minimize the applied bias to minimize the dark etch rate in these processes. By controlling the light intensity, it has been demonstrated that PEC processes can be used to etch a layer of a semiconductor material to a self-limiting thickness by controlling the light intensity (*86*) and operating under short-circuit conditions (*87*).

Solid-phase photochemical reactions using gas-phase reactants also depend on the availability of photogenerated carriers at the surface. Photochemical etching of GaAs by Cl atoms has been shown to be linearly dependent on the concentration of carriers (*88*). Under conditions where the gaseous ambient produces unpinning of the surface Fermi level of GaAs(100), application of a negative bias voltage has been shown to suppress the etching of n^+-GaAs relative to that of *n*- and *p*-GaAs (*89*). Since photocarrier generation requires photons with energy in excess of the bandgap, an extremely high degree of selectivity between chemically similar materials possessing different bandgaps is possible with solid-phase photochemical reactions (*90*). The reaction of Si with Cl has shown that a decrease in the electron concentration increases the rate of the light-induced reaction relative to the dark reaction (*91*). While n^+-Si etches at the same rate in either the dark or the light, *p*-Si etches at a significant rate only under illumination (Fig. 14). Light has also been shown to affect the rate of reaction of Si with F (*92–95*).

3. Selectivity

Material selectivity in laser-induced etching can be divided into two general categories. For processes in the first category, the laser is mainly

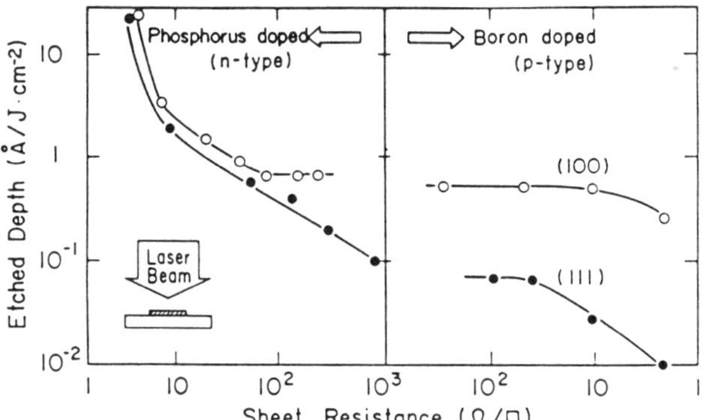

FIG. 14. Etched depth per unit energy versus sheet resistance. [From Arikado *et al.* (*91*).]

responsible for spatial localization of a reaction which can occur in the absence of light under appropriate conditions, and the overall material selectivity is determined by the chemical selectivity of the reaction. Gas-phase photochemical reactions are a prime example of this effect. The principal role of the laser is to photodissociate some reactant precursor molecule. Alternative reactant-generation techniques, such as plasmas, will produce etching with the same chemical selectivity. For example, when light is used to generate Br atoms from HBr (*96*) or CF_3Br (*97*), the same preferential crystallographic etching of GaAs is observed as when a plasma is used to generate Cl or Br, i.e., (111)As > (100) > (110) > (111)Ga. A second example of this category is the localized heating of a material to induce reaction. This is exemplified by the removal of Fe–garnet layers from GGG (*77, 78*), as discussed in Section III,2,d.

Processes in the second category are fundamentally dependent on light to drive the reaction. Solid-phase photochemical reactions of semiconductors using either liquid (wet) or gaseous (dry) reactants are in this category, since these processes require photogeneration of carriers. Since photocarriers are generated only by photons with energies in excess of the bandgap of a semiconductor, it is possible to achieve an extremely high degree of etching selectivity between materials which are chemically very similar but which have different bandgaps. This has been demonstrated by the selective etching of $GaAs_{0.8}P_{0.2}$ under conditions which produce no etching of $GaAs_{0.63}P_{0.37}$ by using photons with energies intermediate between the bandgap energies of the two materials (*90*).

Selective etching based on dopant type has been demonstrated using both wet and dry processes. Selectivity using wet processes arises from galvanic effects, where the light and dark regions of the surface provide anodic and cathodic regions for reaction. In general, illumination enhances etching of *n*-type relative to *p*-type materials. For example, wet etching of GaAs with 257 nm light at 10 W/cm^2 using acid/peroxide gives an etch rate for 10^{18}/cm^3 *n*-GaAs which is 1.5 times higher than for Cr-doped GaAs and 20 times higher than for 10^{18}/cm^3 *p*-GaAs (*72*). With KOH, rates for *n*-GaP which were 10^7 times higher than for *p*-GaP have been reported (*98*). Whereas a hole appears in *n*-GaAs when light locally illuminates the surface, a mesa forms on *p*-GaAs under the same conditions because the light retards the etching of *p*-GaAs relative to the dark etching rate (*99*). The depth or height of these features depends on the relative magnitude of the etch rates in the dark and under illumination. A PEC etching technique for *p*-GaAs has been reported in which an external voltage source was used to switch the potential so photogenerated electrons and majority holes alternately reach the surface (*100*).

In the reaction of dry etchants such as Cl with GaAs, both *n*- and *p*-GaAs etch at the same rate when the substrate is at ground (*88, 89*). Application of an appropriate negative bias has been shown to suppress the etching of 2.5×10^{17}/cm^3 *n*-GaAs almost completely while 1.5×10^{17}/cm^3 *n*-GaAs or *p*-GaAs etch at unimpeded rates under the same conditions, as shown in Figs. 15 and 16 (*89*). Photoluminescence and photoreflectance spectra suggest that this is due to reduction of the electric field in the surface which is necessary to bring photogenerated holes to the surface to drive the photochemical etching reaction.

4. Damage

When relatively low laser power densities are employed, laser-etching processes produce no significant damage to the substrate. However, when very high power densities are used, damage in the form of defects, residual strain, microcracking, surface debris, or changes in morphology can sometimes occur.

Process-induced defects are of special concern in semiconductor materials because of their ability to degrade the electronic properties of the semiconductor material. Melt-based processes produce fewer defects if lower thermal gradients and, therefore, lower recrystallization velocities are involved (*24*). Faster scan speeds produce fewer defects. For example, given a fixed temperature distribution in the material, an increase in scan rate from 1 to 100 cm/sec can decrease defect concentrations in Si by a factor of 100 (*101*). Increasing the speed also narrows the depth distribution of the defects and moves the peak in the distribution toward the surface (*101*).

Generation of point defects in Si using a pulsed Nd–YAG laser requires less power than the threshold power for inducing melting with 532 nm

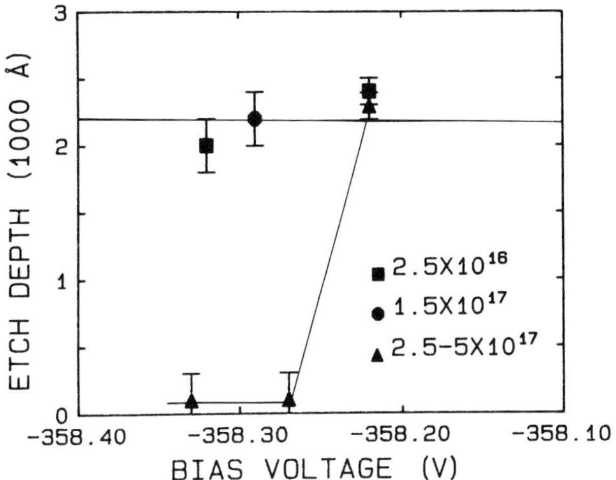

FIG. 15. Plot of etch depth versus applied voltage for three doping levels of *n*-GaAs at 10 W/cm^2 illumination. [From Ashby (*89*).]

light and less than the vaporization threshold energy with 1.06 μm light (*102*). The carrier lifetime is decreased even at power densities of only 60% of the threshold for a visible change in the surface (*102*). The threshold power densities for a 3 nsec pulse of 532 nm light have been shown to be 5.7×10^7, 8.3×10^7, and 1.7×10^8 W/cm^2 for InP, GaAs, and Si,

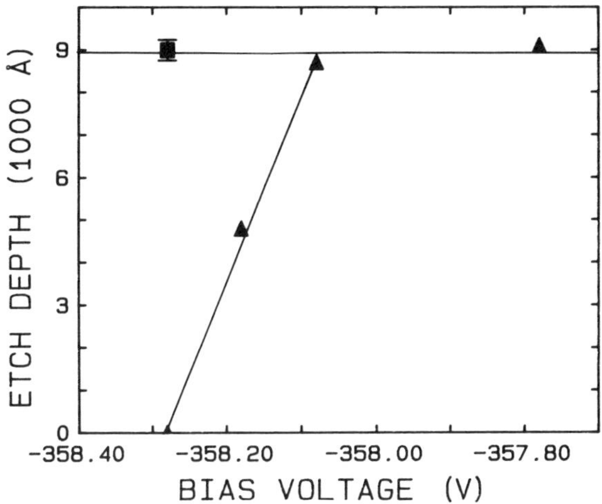

FIG. 16. Plot of etch depth versus applied voltage for 2.5–5 $\times$ 10^{17}/cm^3 *n*-GaAs (▲) and 7–9 $\times$ 10^{18}/cm^3 *p*-GaAs (■) at 1000 W/cm^2 illumination. [From Ashby (*89*).]

respectively (*103*). At these power densities, surface-defect densities on the order of $10^{14}/cm^2$ were produced. Two types of defects are observed in *n*-AlGaAs using 0.5–1.5 MW/cm^2 of cw 647 nm light (*82*), one of which inhibits subsequent etching. At 1.5 MW/cm^2, a large concentration of dark line defects formed. Enhanced recombination at these defects retarded subsequent PEC etching at low laser powers. Damage is more severe with shorter-wavelength light. While 308 nm light does not increase the flat-band voltage for a Si MOS capacitor, 193 nm light produces a shift comparable to that produced by ion-induced damage in reactive ion etching (Fig. 17) (*104*).

Large temperature gradients can produce residual strain and, consequently, lattice defects (*64*). A laser-induced temperature rise of 510 K in GaAs is sufficient to produce a residual strain of 8.6×10^{-3} independent of scan speed (*64*). Repeated shots at power densities well below the single-shot damage threshold can produce residual strain that extends tens of micrometers farther than the original 150 μm beam diameter in Si (*105*).

Large temperature gradients can also produce gross damage such as microfractures in some materials. The thermal stress resistance of a mate-

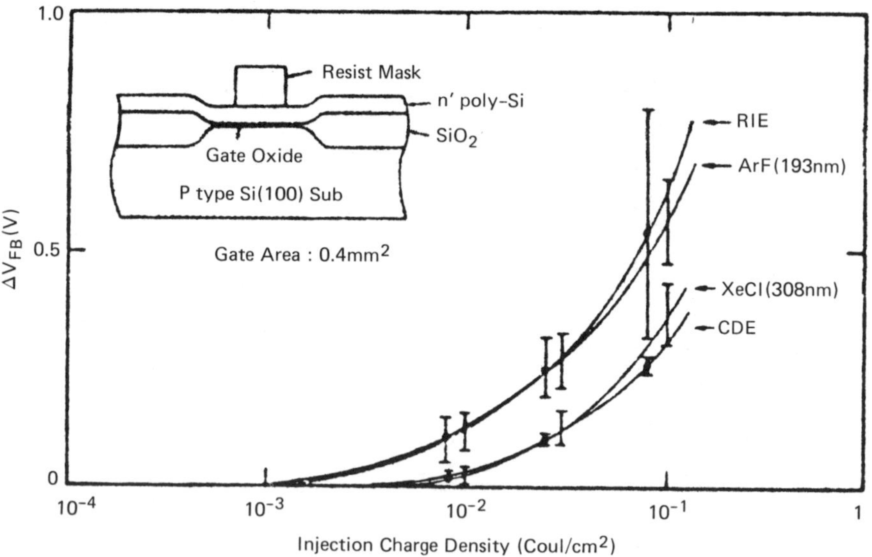

FIG. 17. Flat-band voltage shifts versus injection charge density in the MOS capacitor fabricated by reactive ion etching (RIE), laser etching, and chemical dry etching (CDE) of n^+-poly-Si. The ArF laser was used only for irradiation after poly-Si CDE. [From Itoriike *et al.* (*104*).]

rial provides a guide for predicting whether microcracking will occur. When the ratio of thermal conductivity to the thermal expansion coefficient is small, microfracture is likely (*106*). This problem commonly occurs for ZnO, but not for GaAs, Ge, Si, or SiC. Faceting is seen with Ge when laser pulses with peak power densities greater than 2.5×10^7 W/cm^2 are used (*107*). High-power cw irradiation of the polymer PET, poly(ethylene terephthalate), in air or O_2 to etch vias can produce cracks (*108*).

Ablation processes for materials other than organic polymers almost always leave debris on the surface around the etched feature. In the single-shot removal of metal films, debris on adjacent areas consistent with ejection of molten material is observed (*20*). Even for polymers, debris can be a problem. There is a transition from a rough to smooth surface morphology at a certain absorbed energy density; the required fluence depends on the absorption coefficient at the wavelength employed (*6*). The surface appears to have melted when smooth morphology is observed (*6*). The amount of debris under fixed ablation conditions using 193 nm light varies with polymer type (*109*). Roughened surfaces can be produced in PET (*13*) and PMMA (*4*) with 193 nm light and in polyimide with 308 nm light (*16*).

Low-power laser etching with liquid etchants can produce surface ripples (*110*). This is seen in the etching of GaAs with 530 nm cw light using 1–100 W/cm^2 and an acid/peroxide etching solution. The ripple direction is independent of crystallographic orientation but perpendicular to the laser polarization. The mechanism is unknown.

IV. Applications of Laser-Induced Etching

The important characteristics of the etching processes for different material types are discussed in this section. Tables II–V contain a concise listing of most of the materials etched to date. Several earlier summaries of laser-induced-etching reactions have appeared in the literature (*26, 27, 30, 111, 112*). The many etching processes which require elevated temperatures are also good candidates for laser-induced-etching processes (*113*).

1. Polymers

Laser-induced etching is important for this class of materials principally because of their widespread use as resists in microelectronic device fabrication. The development of single-step processes to remove polymeric

TABLE II

POLYMERS[a]

Two-step photoetching				
	248 nm	193 nm	157 nm	
PMMA[b]	(*8*)	(*50, 51*)	—	
MMA-MAA[c]	—	—	(*39*)	
Low-power one-step photoetching				
	514 nm	257 nm		
PMMA	—	(*14*)		
PET[d]	(*108*)	—		
Nitrocellulose	—	(*14*)		
Photoablation				
	351 nm	308 nm	248 nm	193 nm
PMMA	(*6*)	(*15, 115*)	(*6, 71*)	(*2, 4–6, 10, 15, 71, 109*)
PET	(*6*)	(*15–17*)	(*6, 17*)	(*4, 6, 10, 12, 13, 15, 17*)
Polyimide	(*6, 7*)	(*7, 15–18*)	(*6, 7, 17, 18*)	(*4, 6, 10, 15, 17, 18, 109*)
Polysilane	—	—	(*9, 11*)	—
Polycarbonate	—	—	(*71*)	(*10, 71, 109*)
Nitrocellulose	—	—	—	(*114, 116, 117*)
Cellulose acetate	—	—	—	(*109*)
Polyester	—	—	—	(*109*)
PMIPK[e]	—	—	—	(*109*)
Novalak	—	(*115*)	—	(*109*)
AZ 1350	—	(*16*)	—	—
AZ 2400	—	(*115*)	—	(*109*)

[a] Reference numbers in parentheses.
[b] PMMA: poly(methyl methacrylate).
[c] MMA-MAA: methyl methacrylate–methacrylic acid copolymer.
[d] PET: poly(ethylene terephthalate).
[e] PMIPK: poly(methyl isopropenyl ketone).

resist materials obviates the additional wet-processing step required with lamp-exposed resists. It is this "self-developing" character of the laser-induced process which has made its application to these materials very attractive. The most common method employed is ablative photodecomposition using pulsed uv excimer lasers. UV light is used in preference to visible or IR light because it produces much less thermal damage in areas adjacent to the etched features (*3*). The mechanism involved has been discussed in Section II,1. At subablation laser power densities, nonvolatile residues can form (*14, 114*), presumably due to crosslinking. A similar crosslinking effect is often seen in thermal decomposition of polymers. Although laser-induced etching is generally used in a self-developing mode, lasers are sometimes used at subablation power densities to produce rapid exposure of photoresists. The exposed area is subsequently etched with conventional resist developer solutions (*8, 39, 50*). The laser process requires a lower integrated energy input than a lamp-based process (*8*). The following materials have been etched using laser-induced processes: polymethyl methacrylate (PMMA) (*2, 4–6, 8, 10, 14, 15, 19, 71, 109, 115*), poly(ethylene terephthalate) (PET, Mylar) (*6, 10, 12, 13, 15–17, 108*), polyimide (Kapton) (*6, 7, 10, 15–18, 109*), polycarbonate (*10, 71, 109*), polysilanes (*9, 11*), nitrocellulose (*14, 114, 116, 117*), cellulose acetate (*109*), methyl methacrylate–methacrylic acid copolymer (*39*), polyester (*109*), poly methyl isopropenyl ketone (PMIPK) (*109*), and the resists AZ 2400 (*109, 115*), AZ 1350 (*16, 31, 49*), and Novalak (*109*).

2. Metals

Laser-induced etching of metals has been reported using ablation, thermal, and gas- or liquid-phase photochemical processes (Table III). Ablation has been used directly to remove Ag, Au, Al, Ni, Cu, and Cr films from quartz, glass, and several polymer substrates (*20*) and Al from thermally grown SiO_2 (*118*). In addition, high-power pulses of 308 nm light have been used to ablate or crack the passivating oxide on an Al surface in the presence of Cl_2 to expose the Al surface for reaction. The product $AlCl_x$ was proposed to form during the period between pulses and to be ablated by the next laser pulse (*28*).

A variety of thermal processes have been used to etch metals. Subablation power densities of 248 nm light have been used to crack the Al oxide and permit reaction with Cl_2 (*29*). The laser provides the heat necessary to volatilize the $AlCl_3$ product if the substrate is at room temperature; this is not needed if the substrate has been heated to 350 K. This process has been used to etch part of the Al gate contact in a gate–oxide transistor with no change in the threshold voltage (*29*); this lack of damage to the underlying semiconductor substrate makes this approach more promising than conventional ablation processes for device modification. uv-laser deposition of Zn on an Al surface followed by diffusion of Zn into the Al by heating with a cw visible laser forms a Zn–Al alloy which dissolves much more readily in acetic acid than the Al does (*119*). This microalloying reaction has been used to provide spatially selective etching of Al. A focused cw visible laser has been used to provide thermal control of the balance between Al_2O_3 formation and dissolution in a $H_3PO_4/HNO_3/$

TABLE III

METALS

Material	Wavelength[a]	Mechanism and reactant	Ref.
Ag	308	A[b]	(*20*)
Al	308	A	(*20*)
	580	A	(*118*)
	308	A, TH[c]: Cl_2	(*28*)
	308	TH: Cl_2	(*29*)
	514	TH: $H_3PO_4 + HNO_3 + K_2CR_2O_7$	(*120, 121*)
	514	TH: Zn	(*119*)
Au	308	A	(*20*)
Cr	308	A	(*20*)
Cu	308	A	(*20*)
	308	LPPC[d]: Br_2	(*124*)
	248	LPPC: $Cl^- + Ce^{4+}$	(*124*)
Mo	488, 514	TH: F_2 or O_2	(*63, 67*)
	193	GPPC[e]: NF_3	(*122, 164*)
Ni	308	A	(*20*)
Ta	~10 μm	TH: XeF_2	(*75, 76*)
	~10 μm	GPPC: SF_6	(*75, 76*)
Ti	193	GPPC: NF_3	(*122, 164*)
W	488	TH: F_2	(*63*)
	193	GPPC: F_2CO	(*123, 164*)

[a] Wavelengths in nm unless otherwise noted.
[b] A: photoablation.
[c] TH: thermal.
[d] LPPC: liquid-phase photochemical.
[e] GPPC: gas-phase photochemical.

$K_2Cr_2O_7$ solution (*120, 121*). Both W and Mo have been etched by O_2 and F_2 after heating the metal surface to 700–800 K with a cw visible laser (*63, 67*). In such thermal reactions, heat conduction in the film determines the threshold power for reaction and the feature dimensions (*67*). A thermal process may also be involved in the etching of Ta by XeF_2 with ir radiation from a CO_2 laser (*75, 76*).

Reactant generation by photochemical means has also been used to etch metals. Multiphoton excitation of SF_6 using a CO_2 laser has been used to etch Ta (*75, 76*). Fluorine atoms generated from NF_3 with 193 nm light have been used to etch both Ti and Mo (*122*) and from F_2CO with 193 nm light to etch W (*123*). Copper has been etched with Br atoms generated from Br_2 with 308 nm light (*124*). Copper has also been etched by Cl_2 produced in solution from Cl^- by Ce^{4+}, which in turn was made by photooxidation of Ce^{3+} using 248 nm light (*124*).

3. SEMICONDUCTORS (*124a*)

Ablation processes for etching semiconductors may be undesirable for etching semiconductors for many applications, since the damage usually attendant with these processes degrades the electronic properties of the semiconductor, although laser ablation is currently used commercially for link breaking. Thermal, gas- or liquid-phase photochemical, and solid-phase photochemical processes have also been employed. In many cases, the mechanism of a laser-induced etching process is a combination of these mechanisms (Table IV).

a. Silicon. Silicon has received the most attention in the laser-etching field. Consequently, a variety of different processes have been reported.

TABLE IV

SEMICONDUCTORS

Material	Wavelength[a]	Mechanism and reactant	Ref.
Si	458–514, 694	TH[b]: KOH	(*127, 128*)
	1.06, 10.6 μm	TH: NaOH	(*70*)
	248	TH: Cl + MMA[c]	(*135*)
	458–514	TH: Cl_2	(*62, 69*)
	458–514	TH: HCl	(*62*)
	~10 μm	TH: XeF_2	(*75, 76, 129*)
	~10 μm	GPPC[d]: SF_6	(*75, 76, 131*)
	193	GPPC: F_2CO or NF_3	(*122, 123, 130, 164*)
	193, 308	GPPC: Cl_2 + $Si(CH_3)_4$	(*104*)
	308	GPPC: Cl_2	(*91*)
	550	SPPC–PEC[e]: HF	(*85*)
	458–514	SPPC: XeF_2	(*92–95*)
	350–647	SPPC: CF_4 + O_2 plasma	(*35, 79, 81, 133, 134*)
	350–647	SPPC: NF_3 plasma	(*35*)
Ge	488, 514	GPPC: Br_2	(*107, 136, 137*)
GaAs	458–514	TH: CCl_4	(*64–66, 140*)
	532	TH: CCl_4 + H_2	(*139*)
	514	TH: CCl_4, $SiCl_4$, or $GeCl_4$	(*61*)
	514	TH: Cl_2	(*140*)
	488	GPPC + TH: Cl_2	(*138*)
	193	GPPC + TH: HBr	(*96*)
	193, 351	GPPC + TH: CF_3Br or CH_3Br	(*97, 141*)
	257	GPPC: CF_3I or CH_3Br	(*53, 143*)
n-GaAs	413, 521, 633	LPPC[f]: I_2 or Br_2 in H_2O	(*55*)
GaAs	visible	SPPC–PEC	(*84*)
n-GaAs	visible	SPPC–PEC: KOH	(*87*)

(*Continues*)

TABLE IV (*Continued*)

Material	Wavelength[a]	Mechanism and reactant	Ref.
n-GaAs	visible	SPPC–PEC: NaOH + EDTA	(*83*)
p-GaAs	633	SPPC–PEC: H_2SO_4 + NaSCN	(*100*)
GaAs	248, 257	SPPC: HNO_3	(*68*)
n- or SI-GaAs	334, 514, 633	SPPC: HNO_3 or KOH	(*80, 144, 145*)
p-, *n*-, or SI[g]-GaAs	257, 458, 488, 514	SPPC: H_2SO_4 + H_2O_2	(*46, 58, 72, 99, 145*)
n-GaAs	visible	SPPC: $Fe_2(SO_4)_3$ + EDTA	(*147*)
GaAs	647	SPPC: Zn	(*148*)
p- or *n*-GaAs	488, 514, 766	SPPC: Cl	(*88–90*)
Ga(As, P)	514, 766	SPPC: Cl	(*90*)
GaP	351	SPPC + TH: KOH	(*98*)
(Al, Ga)As	647	SPPC: H_3PO_4, H_2O_2 in CH_3OH	(*47, 82*)
(Ga, In)P	248	GPPC: CH_3X + H_2	(*150*)
(Ga, In)As	visible	SPPC–PEC	(*84*)
(Ga, In)(As, P)	248	GPPC: CH_3X + H_2	(*150*)
	442, 488, 633	SPPC–PEC: HF or KOH	(*56, 84*)
InP	514	TH: CCl_4, $SiCl_4$, or $GeCl_4$	(*149*)
	248, 257	GPPC: CH_3I, CH_3Br, or CF_3I	(*53, 143, 150*)
n-InP	442, 488, 633	SPPC–PEC: HF, HCl, HBr, H_2SO_4, or KOH	(*38, 56, 57, 84*)
SI-InP	488, 514	SPPC: HCl + HNO_3	(*145*)
InSb	514	TH: CCl_4, $SiCl_4$, or $GeCl_4$	(*149*)
ZnSe	458, 473, 488	SPPC: HNO_3 + HCl	(*152*)
CdS	488	SPPC: H_2SO_4 + H_2O_2	(*46, 145*)
	458, 488	SPPC: HNO_3	(*153*)
	458, 473, 488	SPPC: HNO_3 + HCl	(*152*)
Cd(S, Se)	633	SPPC: HCl	(*154*)
CdSe	350–1100	SPPC-PEC: KCl	(*41*)
CdTe	458–514	TH	(*74, 151*)

[a] Wavelengths in nm unless noted.
[b] TH: thermal.
[c] GPPC: gas-phase photochemical.
[d] LPPC: liquid-phase photochemical.
[e] SPPC: solid-phase photochemical; PEC: photoelectrochemical.
[f] MMA: methyl(methacrylate).
[g] SI: semi-insulating.

Poly-Si link breaking by laser ablation is well established in commercial processing for circuit restructuring (*125, 126*) and will not be discussed further here. The fastest chemical processes are primarily thermal in nature. Etching rates of 10 to 20 μm/sec have been reported when the laser melts the silicon surface in the presence of a reactive gas or solution. While Si(100) and Si(111) etch at the same rate (*127*), careful control of laser intensity permits selective etching of amorphous Si relative to crystalline Si (*128*). When heated to temperatures just below the melting point, preferential crystallographic etching occurs but *n*-Si and *p*-Si etch identically (*69*). By selecting the IR wavelength employed, it is possible to produce etching either by heating the solution in the beam path or by heating the Si surface (*70*). If one regards lattice vibrational excitation as a thermal process, the IR-induced etching of Si by XeF_2 would be in the thermal-etching category (*75, 129*).

Gas-phase and liquid-phase photochemical processes generally employ beams which are perpendicular to the substrate surface and, therefore, frequently have a thermal component to their mechanism. Most gas-phase processes are based on generation of reactive halogen species: F from F_2CO or NF_3 with 193 nm light (*122, 123, 130*), Cl from Cl_2 with 308 nm pulsed (*91, 104*) or 514 nm cw light (*62*), and (SF_5 + F) (*131*) or vibrationally excited SF_6 (*75, 76, 131, 132*) from SF_6 with multiphoton absorption of 10 μm light.

Solid-phase photochemical mechanisms frequently manifest their importance in a process by a difference in etching rates between *n*-type and *p*-type Si or by differences between materials with different carrier concentrations. A wavelength dependence can also be indicative. A pronounced difference is seen in the etching of *n*- and *p*-Si by Cl atoms generated from Cl_2 with 308 nm light (*91*). While *n*-Si etches without direct illumination of the surface, undoped and *p*-Si require incident light to etch. A crystallographic dependence of the etch rate is seen for lightly doped *n*-Si but not for heavily doped *n*-Si (*91*). The role of surface electronic excitation is shown in the etching of Si by XeF_2 under visible light by a difference in the volatile-reaction-product distribution relative to the dark thermal reaction (*92, 93*) and a difference in the photoreaction product distribution for *n*- and *p*-Si (*94, 95*). The role of photogenerated carriers in the PEC processes, such as the etching of *n*-Si by HF (*85*), is clearly evident in the rate dependence on minority-carrier flow to the surface.

Laser-enhanced plasma etching of Si is possible due to both thermal (*133–135*) and solid-phase photochemical (*35, 79, 81*) mechanisms. The role of photogenerated carriers in reactions under cw illumination is clearly shown by the higher reaction rates obtained using shorter-wavelength light of moderate intensity when $10^{20}/cm^3$ Si of either *n*- or *p*-type is etched (*35, 79, 81*). At higher intensities, thermal factors can override photochemical contributions to the etching process (*81*). A similar wavelength dependence would be expected in nonplasma etching of Si as well.

b. Germanium. Most of the very early laser-induced etching work was performed using halogens to etch Ge. Lasers have been used both to photodissociate Br_2 (*136, 137*) and to heat the substrate surface to temperatures in excess of 1000 K to produce rapid etching (*107*).

c. GaAs, Ga(As, P), and GaP. Gallium arsenide has been second only to Si in the level of interest and effort in laser-induced etching. The processes are generally based on thermal, gas- or liquid-phase photochemical, and solid-phase photochemical mechanisms. The fastest rates are achieved by driving thermal reactions by laser-induced heating. Rates up to 33 μm/sec have been reported when 488 nm light was used to heat GaAs to melting or near-melting while also photogenerating Cl atoms from Cl_2 (*138*). Laser-enhanced reactive ion etching (RIE) using CCl_4 and H_2 gives rates up to 10 times faster than RIE alone when GaAs is heated to near the melt (*139*). At submelting temperatures, rates up to 5 μm/sec using scanning 514 nm beams have been reported using CCl_4 (*61, 64–66, 140*) or Cl_2 (*140*). Residual strain from the high temperature gradients is possible with these thermal processes (*64*). The temperature rise during these pyrolytic etching processes has been studied using $SiCl_4$, CCl_4, and $GeCl_4$ (*61*).

Several gas-phase photochemical processes in which uv excimer lasers are used to generate halogen atoms often have significant thermal components to the overall reaction rate (*96, 97, 141, 142*) but can also etch GaAs at low laser power densities which produce negligible temperature increases (*53*). These often show preferential crystallographic etching rates similar to non-laser-enhanced thermal reactions (*96, 97, 141*). Amorphous GaAs etches at a faster rate than crystalline GaAs (*143*). Irradiation using power densities sufficient to produce ablation produces etching in the presence of reactive gases such as CF_3Br and CH_3Br that is 100 times faster than ablation alone (*97*). An XPS study of the etching of GaAs by HCl using 193 nm light has been reported (*54*).

Solid-phase photochemical processes can be divided into PEC processes requiring an external voltage supply and current flow and those processes which do not. Most PEC processes are used to etch *n*-GaAs (*83, 86, 87*), but bias potential cycling has been used to overcome the stability of *p*-GaAs toward reductive decomposition by alternately bringing holes and electrons to the surface (*100*). Preferential crystallographic etching is not present in PEC etching if sufficient light intensity is used (*83*). PEC has been used to self-limit the thickness of *n*-GaAs on semi-insulating (SI) GaAs by controlling the light intensity and the applied bias (*86, 87*).

Most non-PEC laser-assisted wet etching of GaAs has employed acid solutions (*46, 68, 80, 144*) or acid/peroxide solutions (*45–47, 58, 72, 99, 145*); etching of GaAs (*145, 146*) and GaP (*98*) using KOH is also possible. A substantial dark etch rate occurs if much H_2O_2 is used since acid/peroxide is a standard etching solution for nonphotochemical etching of GaAs. Light can, in essence, replace the chemical oxidant, H_2O_2, in photochemical etching. At lower power densities, uv light produces faster etching than visible light (*72, 80, 144*), but rates are relatively independent of wavelength at higher powers (*72*). Rates for *n*-GaAs are somewhat faster than for SI-GaAs and 10 to 100 times faster than for *p*-GaAs (*72*). A process combining both photogeneration of Br or I in an aqueous solution of Br_2 or I_2 and photocarrier generation has been reported (*55*). This process etches *n*-GaAs but not SI-GaAs. A process using an Fe^{III}/EDTA solution and visible light gives doping-selective etching with $n^- > n^+ >> p^+ = p$ (*147*).

A solid-phase photochemical process for etching GaAs (*88*) and Ga(As, P) (*90*) with Cl atoms exhibits no measurable dark etching under appropriate pressure and flow conditions. The suppression of the etching of n^+-GaAs relative to *n*-GaAs or *p*-GaAs by applying a bias voltage requires less than a factor of three difference in doping level to give greater than a 20-fold difference in rate (*89*). The process has been demonstrated to produce virtually total selectivity between Ga(As, P) materials of different As-to-P ratios with photons of energies greater than the bandgap energy of the material with lower P content but lower than the bandgap energy of the material with higher P content (*90*). A solid-phase photochemical reaction of GaAs in a Zn atmosphere has been reported (*148*). The reaction of Zn with GaAs which is driven by light produces more volatile Zn_3As_2 and Ga.

d. Other III-V Semiconductors. Both InP and InSb have been etched by CCl_4, $GeCl_4$, and $SiCl_4$ in a rapid thermal reaction following substrate heating with a laser (*149*). Both 257 nm cw (*53*) and 248 nm pulsed (*150*) light can be used to photodissociate methyl halides to etch InP (*53, 150*), GaInAsP (*150*), and GaInP (*150*). Simultaneous heating of the surface enhances the reaction rate (*53*), as does addition of H_2 to reduce the formation of surface oxides (*150*). Laser-PEC etching of InP, InGaAs, and InGaAsP has been reported using both acidic solutions (*56, 57, 84*) and basic solutions (*56*). Addition of H_2O_2 to acidic solutions has little if any effect on the PEC etching rate (*57*). Non-PEC etching has also been reported for AlGaAs using acid/peroxide solutions (*47, 82*). In general, one may expect that processes which have been developed for GaAs may work with only slight modification for other III–V materials.

e. II–VI Semiconductors. Cadmium telluride has been etched with cw visible light by a photosublimation mechanism which is believed to combine both thermal and photoinduced carrier effects (*74, 151*). Both CdS and ZnSe have been etched by a PEC process using $HNO_3/HCl/H_2O$ that shows some wavelength dependence of both the etch rate and the surface etch-pit density (*152*). PEC etching using a KCl solution has also been reported for *n*-CdSe (*41*). Non-PEC etching of CdS using acid or acid/peroxide solutions (*46, 145, 153*) and of Cd(S, Se) using HCl in methanol (*154*) also have been demonstrated.

4. Other Inorganic Materials

Laser etching of Si_3N_4 by F or Br generated by gas-phase photolysis of molecular precursors has been reported (*130, 155*) (Table V). Photoablation of SiO_x using 248 nm pulses has been used to pattern an etching mask for plasma etching of Si (*156*). Several studies of laser-induced etching of CVD or plasma-grown SiO_2 on Si have been reported. Some involve simply heating the material with the laser beam (*157*), but most have used the laser to generate reactive gas-phase species such as CF_2 (*123, 130, 158*), CF_3 and Br (*155, 158*), or F (*54, 159*). Higher rates are achieved when the beam is perpendicular rather than parallel to the surface, resulting in a thermal contribution to the reaction (*158, 159*). Fused silica has been etched using ablation (*160*), ablation in the presence of a reactive gas (*160*), and thermally (*157*). Pyrex has been etched using the laser to enhance a thermal reaction, (*157*) and borosilicate glass has been etched using laser generation of CF_2 (*161*).

Laser heating has been used to increase the reaction of sapphire with plasma-generated reactants (*157*). Laser-induced melting of an alumina/TiC ceramic has produced very rapid etching by KOH (*127*). Both TiC and TiB_2 have been etched in a few torr of Cl_2 during heating with a laser (*162*). The ablation of MnZn–ferrite is two times faster in the presence of a reactive gas such as CCl_4, and the presence of the gas prevents the surface cracking seen with ablation alone (*93*). Selective etching of $(YBi)_3(FeGa)_5O_{12}$ or $(BiGdLu)_3(FeGa)_5O_{12}$ from a $Gd_3Ga_5O_{12}$ substrate is possible using a laser of appropriate wavelength to selectively heat the Fe–garnet layer to drive its thermal reaction with H_3PO_4 (*77, 78*). The fusion reaction of molten $LiNbO_3$ with KF to produce highly soluble niobium oxyfluoride salts has been used to pattern $LiNbO_3$ by using a laser to produce a localized molten region on the substrate surface (*163*).

TABLE V

OTHER INORGANIC MATERIALS

Material	Wavelength[a]	Mechanism and reactant	Ref.
Fused SiO_2	10.6 μm	A[b] in presence of HCl	(*160*)
	248, 351	TH[c]: SiH_4	(*157*)
Quartz	193	GPPC[d]: $NF_3 + H_2$	(*54*)
SiO_2 on quartz	10.6 μm	GPPC: CF_3Br	(*155*)
	193, 248; 9.2–9.3 μm	GPPC: CF_3Br or CF_3I	(*158*)
	193, 248; 9.2 μm	GPPC: CF_2Br_2	(*158*)
	193	GPPC: C_2F_4	(*158*)
SiO_2 on Si	458	TH: Cl_2	(*76*)
	248, 351	TH: SiH_4	(*157*)
	193	GPPC: $NF_3 + H_2$	(*54, 159*)
SiO_x	248	A	(*156*)
Pyrex	193	TH: H_2	(*31*)
	248, 351	TH: SiH_4	(*157*)
Borosilicate glass	193, 248	GPPC: CF_2Br_2	(*161*)
Si_3N_4	10.6 μm	GPPC: CF_3Br	(*155*)
	193	GPPC: F_2CO or NF_3	(*130*)
Alumina	248, 351	TH: SiH_4	(*157*)
Alumina/TiC	514	TH: KOH	(*127*)
TiC	514	TH: Cl_2	(*162*)
TiB_2	514	TH: Cl_2	(*162*)
MnZn–ferrite	458–514	TH: CCl_4	(*93*)
Fe–garnet	520, 581	TH: H_3PO_4	(*77, 78*)
$LiNbO_3$	248, 270	TH: KF	(*163*)

[a] Wavelengths in nm unless otherwise noted.
[b] A: photoablation.
[c] TH: thermal.
[d] GPPC: gas-phase photochemical.

V. Concluding Remarks

Much progress has been made in the past decade in the development of laser-induced etching processes which are applicable to a wide variety of thin-film materials. The possibilities for further development of existing processes and for identification of new processes to provide spatially localized, high-resolution, material-selective etching of important thin-film materials make the laser-based approach to etching an area where one may expect to continue to see exciting new developments in the years ahead.

ACKNOWLEDGMENTS

The author wishes to thank J. Y. Tsao for his critical reading of the manuscript. This work was performed at Sandia National Laboratories, supported by the U.S. Department of Energy.

References

1. T. J. Chuang, *J. Vac. Sci. Technol.* **21,** 798 (1982).
2. B. J. Garrison and R. Srinivasan, *Appl. Phys. Lett.* **44,** 849 (1984).
3. B. J. Garrison and R. Srinivasan, *J. Appl. Phys.* **57,** 2909 (1985).
4. R. Srinivasan, *J. Vac. Sci. Technol., B*[2]**1,** 923 (1983).
5. G. M. Davis, M. C. Gower, C. Fotakis, T. Efthimiopoulos, and P. Argyrakis, *Appl. Phys.* [*Part*]*A* **A36,** 27 (1985).
6. G. Koren and J. T. C. Yeh, *J. Appl. Phys.* **56,** 2120 (1984).
7. J. H. Brannon, J. R. Lankard, A. I. Baise, F. Burns, and J. Kaufman, *J. Appl. Phys.* **58,** 2036 (1985).
8. Y. Kawamura, K. Toyoda, and S. Namba, *Appl. Phys. Lett.* **40,** 374 (1982).
9. A. W. Johnson, J. M. Ziegler, M. E. Riley, and L. A. Harrah, *in* "Beam-Induced Chemical Processes, Extend Abstracts, 1985 Fall Meeting of the Materials Research Society" (R. J. von Gutfeld, J. E. Greene, and H. Schlossberg, eds.), p. 87. Materials Research Society, Pittsburgh, Pennsylvania, 1985.
10. R. Srinivasan and B. Braren, *J. Polym. Sci., Polym. Chem. Ed.* **22,** 2601 (1984).
11. J. M. Ziegler, L. A. Harrah, and A. W. Johnson, *Proc. SPIE—Int. Soc. Opt. Eng.* **539,** 166 (1985).
12. R. Srinivasan and W. J. Leigh, *J. Am. Chem. Soc.* **104,** 6784 (1982).
13. R. Srinivasan and V. Mayne-Banton, *Appl. Phys. Lett.* **41,** 576 (1982).
14. J. E. Bjorkholm, L. Eichner, J. C. White, R. E. Howard, and H. G. Craighead, *J. Appl. Phys.* **58,** 2098 (1985).
15. P. E. Dyer and R. Srinivasan, *Appl. Phys. Lett.* **48,** 445 (1986).
16. J. E. Andrew, P. E. Dyer, D. Forster, and P. H. Key, *Appl. Phys. Lett.* **43,** 717 (1983).
17. P. E. Dyer and J. Sidhu, *J. Appl. Phys.* **57,** 1420 (1985).
18. G. Gorodetsky, T. G. Kazyaka, R. L. Melcher, and R. Srinivasan, *Appl. Phys. Lett.* **46,** 828 (1985).
19. H. H. G. Jellinek and R. Srinivasan, *J. Phys. Chem.* **88,** 3048 (1984).
20. J. E. Andrew, P. E. Dyer, R. E. Greenough, and P. H. Key, *Appl. Phys. Lett.* **43,** 1046 (1983).
21. M. Lax, *J. Appl. Phys.* **48,** 3919 (1977).
22. M. Lax, *Appl. Phys. Lett.* **33,** 786 (1978).
23. E. Liarokapis and Y. S. Raptis, *J. Appl. Phys.* **57,** 5123 (1985).
24. P. Schvan and R. E. Thomas, *J. Appl. Phys.* **57,** 4738 (1985).
25. D. L. Kwong and D. M. Kim, *J. Appl. Phys.* **54,** 366 (1983).
26. J. F. Ready, "Effects of High-Power Laser Radiation." Academic Press, New York, 1971.
27. Y. Rytz-Froidevaux, R. P. Salathe, and H. H. Gilgen, *Appl. Phys.* [*Part*]*A* **A37,** 121 (1985).
28. G. Koren, F. Ho, and J. J. Ritsko, *Appl. Phys. Lett.* **46,** 1006 (1985).

29. K. E. Greenberg, A. W. Johnson, J. W. Medernach, and K. Jungling, *in* "Beam-Induced Chemical Processes, Extend Abstracts, 1985 Fall Meeting of the Materials Research Society" (R. J. von Gutfeld, J. E. Greene, and H. Schlossberg, eds.), p. 59. Materials Research Society, Pittsburgh, Pennsylvania, 1985.
30. D. J. Ehrlich and J. Y. Tsao, *J. Vac. Sci. Technol., B*[2]**1,** 969 (1983).
31. D. J. Ehrlich, J. Y. Tsao, and C. O. Bozler, *J. Vac. Sci. Technol., B*[2]**3,** 1 (1985).
32. T. R. Loree, R. C. Sze, D. L. Barker, and P. B. Scott, *IEEE J. Quantum Electron.* **QE-15,** 337 (1979).
33. C. Hill, *in* "Laser Annealing of Semiconductors" (J. M. Poate and J. W. Mayer, eds.), p. 479. Academic Press, New York, 1982.
34. A. G. Cullis, H. C. Webber, and P. Bailey, *J. Phys. E.* **12,** 688 (1979).
35. G. M. Reksten, W. Holber, and R. M. Osgood, Jr., *Appl. Phys. Lett.* **48,** 551 (1986).
36. K. Jain, C. G. Willson, and B. J. Lin, *IBM J. Res. Dev.* **26,** 151 (1982).
37. G. L. Loper and M. D. Tabat, *Proc. SPIE—Int. Soc. Opt. Eng.* **621,** 87 (1986).
38. J. Cheng and P. A. Kohl, *in* "Laser-Controlled Chemical Processing of Surfaces' (A. W. Johnson, D. J. Ehrlich, and H. R. Schlossberg, eds.), p. 127. Am. Elsevier, New York, 1984.
39. H. G. Craighead, J. C. White, R. E. Howard, L. D. Jackel, R. E. Behringer, J. E. Sweeney, and R. W. Epworth, *J. Vac. Sci. Technol., B*[2]**1,** 1186 (1983).
40. G. L. Loper and M. D. Tabat, *J. Appl. Phys.* **58,** 3649 (1985).
41. C. A. Kavassalis, D. H. Longendorfer, R. A. LeLievre, and R. D. Rauh, *in* "Laser-Controlled Chemical Processing of Surfaces" (A. W. Johnson, D. J. Ehrlich, and H. R. Schlossberg, eds.), p. 151. Am. Elsevier, New York, 1984.
42. D. J. Ehrlich and J. Y. Tsao, *Appl. Phys. Lett.* **44,** 267 (1984).
43. L. V. Belyakov, D. N. Goryachev, L. G. Paritskii, S. M. Ryvkin, and O. M. Sreseli, *Sov. Phys.—Semicond.* **10,** 678 (1976).
44. Zh. I. Alferov, D. N. Goryachev, S. A. Gurevich, M. N. Mizerov, E. L. Portnoi, and B. S. Ryvkin, *Sov. Phys.—Tech. Phys.* (*Engl. Transl.*) **21,** 857 (1976).
45. D. V. Podlesnik, H. H. Gilgen, R. M. Osgood, Jr., and A. Sanchez, *Appl. Phys. Lett.* **43,** 1083 (1983).
46. D. V. Podlesnik, H. H. Gilgen, R. M. Osgood, A. Sanchez, and V. Daneu, *in* "Laser Diagnostics and Photochemical Processing for Semiconductor Devices" (R. M. Osgood, S. R. J. Brueck, and H. R. Schlossberg, eds.), p. 57. Am. Elsevier, New York, 1983.
47. L. V. Belyakov, D. N. Goryachev, M. N. Mizerov, and E. L. Portnoi, *Sov. Phys.—Tech. Phys.* (*Engl. Transl.*) **19,** 837 (1974).
48. H. J. Lezec, E. H. Anderson, and H. I. Smith, *J. Vac. Sci. Technol., B*[2]**1,** 1204 (1983).
49. L. V. Belyakov, D. N. Goryachev, and O. M. Sreseli, *Sov. Phys.—Tech. Phys.* (*Engl. Transl.*) **24,** 511 (1979).
50. A. M. Hawryluk, H. I. Smith, and D. J. Ehrlich, *J. Vac. Sci. Technol., B*[2]**1,** 1200 (1983).
51. A. M. Hawryluk, H. I. Smith, R. M. Osgood, and D. J. Ehrlich, *Opt. Lett.* **7,** 402 (1982).
52. A. Roth, "Vacuum Technology," p. 36. North-Holland, Publ., Amsterdam, 1976.
53. D. J. Ehrlich, R. M. Osgood, Jr., and T. F. Deutsch, *IEEE J. Quantum Electron.* **QE-16,** 1233 (1980).
54. M. Hirose, S. Yokoyama, and Y. Yamakage, *J. Vac. Sci. Technol., B*[2]**3,** 1445 (1985).

55. R. W. Haynes, G. M. Metze, V. G. Kreismanis, and L. F. Eastman, *Appl. Phys. Lett.* **37,** 344 (1980).
56. R. M. Lum, A. M. Glass, F. W. Ostermayer, Jr., P. A. Kohl, A. A. Ballman, and R. A. Logan, *J. Appl. Phys.* **57,** 39 (1985).
57. R. M. Lum, F. W. Ostermayer, Jr., P. A. Kohl, A. M. Glass, and A. A. Ballman, *Appl. Phys. Lett.* **47,** 269 (1985).
58. D. V. Podlesnik, H. H. Gilgen, and R. M. Osgood, Jr., *in* "Laser Chemical Processing of Semiconductor Devices, Extend Abstracts, 1984 Fall Meeting of the Materials Research Society" (F. A. Houle, T. F. Deutsch, and R. M. Osgood, Jr., eds.), p. 161. Materials Research Society, Pittsburgh, Pennsylvania, 1984.
59. R. F. Wood and G. E. Giles, *Phys. Rev. B: Condens. Matter* [3] **23,** 2923 (1981).
60. P. Baeri, S. U. Campisano, G. Foti, and E. Rimini, *J. Appl. Phys.* **50,** 788 (1979).
61. M. Takai, H. Nakai, J. Tsuchimoto, K. Gamo, and S. Namba, *Jpn. J. Appl. Phys.* **24,** L705 (1985).
62. D. J. Ehrlich, R. M. Osgood, Jr., and T. F. Deutsch, *App. Phys. Lett.* **38,** 1018 (1981).
63. G. Koren, *Appl. Phys. Lett.* **47,** 1012 (1982).
64. M. Takai, H. Nakai, S. Nakashima, T. Minamisono, K. Gamo, and S. Namba, *Jpn. J. Appl. Phys.* **24,** L755 (1985).
65. M. Takai, J. Tokuda, H. Nakai, K. Gamo, and S. Namba, *Jpn. J. Appl. Phys.* **22,** L757 (1983).
66. M. Takai, J. Tsuchimoto, H. Nakai, K. Gamo, and S. Namba, *Jpn. J. Appl. Phys.* **23,** L852 (1984).
67. G. Koren, *J. Appl. Phys. 59,* 1667 (1986).
68. D. V. Podlesnik, H. H. Gilgen, P. D. Brewer, D. M. McClure, and R. M. Osgood, Jr., *in* "Laser Chemical Processing of Semiconductor Devices, Extend Abstracts, 1984 Fall Meeting of the Materials Research Society" (F. A. Houle, T. F. Deutsch, and R. M. Osgood, Jr., eds.), p. 109. Materials Research Society, Pittsburgh, Pennsylvania, 1984.
69. D. J. Ehrlich, D. J. Silversmith, R. W. Mountain, and J. Tsao, *IEEE Trans. Components, Hybrids, Manuf. Technol.* **CHMT-5,** 520 (1982).
70. F. V. Bunkin, B. S. Luk'yanchuk, G. A. Shafeev, E. K. Kozlova, A. I. Portniagin, A. A. Yeryomenko, P. Mogyorosi, and J. G. Kiss, *Appl. Phys.* [*Part*] A **A37,** 117 (1985).
71. B. Braren and R. Srinivasan, *J. Vac. Sci. Technol., B*[2]**3,** 913 (1985).
72. D. V. Podlesnik, H. H. Gilgen, and R. M. Osgood, Jr., *Appl. Phys. Lett.* **45,** 563 (1984).
73. D. V. Podlesnik, H. H. Gilgen, and R. M. Osgood, Jr., *Appl. Phys. Lett.* **48,** 496 (1986).
74. C. Arnone, M. Rothschild, and D. J. Ehrlich, *Appl. Phys. Lett.* **48,** 736 (1986).
75. T. J. Chuang, *J. Vac. Sci. Technol.* **18,** 638 (1981).
76. F. A. Houle and T. J. Chuang, *J. Vac. Sci. Technol.* **20,** 790 (1982).
77. K. Ando and S. Tsukahara, *Jpn. J. Appl. Phys.* **21,** L347 (1982).
78. K. Ando, N. Takeda, and N. Koshizuka, *Appl. Phys. Lett.* **46,** 1107 (1985).
79. G. Reksten, W. Holber, and R. M. Osgood, Jr., *in* "Laser Chemical Processing of Semiconductor Devices, Extend Abstracts, 1984 Fall Meeting of the Materials Research Society" (F. A. Houle, T. F. Deutsch, and R. M. Osgood, Jr., eds.), p. 115. Materials Research Society, Pittsburgh, Pennsylvania, 1984.
80. G. C. Tisone and A. W. Johnson, *Appl. Phys. Lett.* **42,** 530 (1983).
81. W. Holber, G. Reksten, and R. M. Osgood, Jr., *Appl. Phys. Lett.* **46,** 201 (1985).
82. B. Zysset and R. P. Salathe, *Appl. Phys. Lett.* **45,** 428 (1984).
83. A. Yamamoto and S. Yano, *J. Electrochem. Soc.* **122,** 260 (1975).

84. K. D. Cummings, L. R. Harriott, G. C. Chi, and F. W. Ostermayer, Jr., *Appl. Phys. Lett.* **48,** 659 (1986).
85. H. J. Hoffman and J. M. Woodall, *Appl. Phys.* [*Part*] *A* **A33,** 243 (1984).
86. A. Shimano, H. Takagi, and G. Kano, *IEEE Trans. Electron. Devices* **ED-26,** 1690 (1979).
87. H. J. Hoffman, J. M. Woodall, and T. I. Chappell, *Appl. Phys. Lett.* **38,** 564 (1981).
88. C. I. H. Ashby, *Appl. Phys. Lett.* **45,** 892 (1984).
89. C. I. H. Ashby, *Appl. Phys. Lett.* **46,** 752 (1985).
90. C. I. H. Ashby and R. M. Biefeld, *Appl. Phys. Lett.* **47,** 62 (1985).
91. T. Arikado, M. Sekine, H. Okano, and Y. Horiike, *in* "Laser-Controlled Chemical Processing of Surfaces" (A. W. Johnson, D. J. Ehrlich, and H. R. Schlossberg, eds.), p. 167. Am. Elsevier, New York, 1984.
92. F. A. Houle, *Chem. Phys. Lett.* **95,** 5 (1983).
93. F. A. Houle, *Proc. SPIE—Int. Soc. Opt. Eng.* **385,** 127 (1983).
94. F. A. Houle, *in* "Laser-Controlled Chemical Processing of Surfaces" (A. W. Johnson, D. J. Ehrlich, and H. R. Schlossberg, eds.), p. 203. Am. Elsevier, New York, 1984.
95. F. A. Houle, *J. Chem. Phys.* **80,** 485 (1984).
96. P. D. Brewer, D. McClure, and R. M. Osgood, Jr., *Appl. Phys. Lett.* **47,** 310 (1985).
97. P. Brewer, S. Halle, and R. M. Osgood, Jr., *Appl. Phys. Lett.* **45,** 475 (1984).
98. A. W. Johnson and G. C. Tisone, *in* "Laser-Controlled Chemical Processing of Surfaces" (A. W. Johnson, D. J. Ehrlich, and H. R. Schlossberg, eds.), p. 145. Am. Elsevier, New York, 1984.
99. F. Kuhn-Kuhnenfeld, *J. Electrochem. Soc.* **119,** 1063 (1972).
100. F. W. Ostermayer, Jr. and P. A. Kohl, *Appl. Phys. Lett.* **39,** 76 (1981).
101. A Chantre, M. Kechouane, G. Auvert, and D. Bois, *Appl. Phys. Lett.* **43,** 98 (1983).
102. D. L. Parker, *in* "Energy Beam-Solid Interactions and Transient Thermal Processing" (J. C. C. Fan and N. M. Johnson, eds.), p. 359. Am. Elsevier, New York, 1984.
103. J. M. Moison and M. Bensoussan, *Appl. Surf. Sci.* **20,** 84 (1984).
104. Y. Horiike, M. Sekine, K. Horioka, T. Arikado, M. Nakase, and H. Okano, *in* "Laser Chemical Processing of Semiconductor Devices, Extend Abstracts, 1984 Fall Meeting of the Materials Research Society" (F. A. Houle, T. F. Deutsch, and R. M. Osgood, Jr., eds.), p. 99. Materials Research Society, Pittsburgh, Pennsylvania, 1984.
105. P. M. Fauchet, I. H. Campbell, and F. Adar, *Appl. Phys. Lett.* **47,** 479 (1985).
106. R. F. Wood and D. H. Lowndes, *Cryst. Lattice Defects Amorphous Mater.* **12,** 475 (1985).
107. G. P. Davis, C. A. Moore, and R. A. Gottscho, *Proc. SPIE—Int. Soc. Opt. Eng.* **459,** 115 (1984).
108. T. J. Chuang *in* "Laser Diagnostics and Photochemical Processing for Semiconductor Devices" (R. M. Osgood, S. R. J. Brueck, and H. R. Schlossberg, eds.), p. 45. Am. Elsevier, New York, 1983.
109. S. Rice and K. Jain, *Appl. Phys.* [*Part*] *A* **A33,** 195 (1984).
110. N. Tsukada, S. Sugata, H. Saito, and Y. Mita, *Appl. Phys. Lett.* **43,** 189 (1983).
111. R. M. Osgood, Jr., *Annu. Rev. Phys. Chem.* **34,** 77 (1983).
112. F. A. Houle, *Proc. SPIE—Int. Soc. Opt. Eng.* **459,** 110 (1984).
113. W. Kern and C. A. Deckert, *in* "Thin Film Processes" (J. L. Vossen and W. Kern, eds.), p. 401. Academic Press, New York, 1978.
114. M. W. Geis, J. N. Randall, T. F. Deutsch, N. N. Efremow, J. P. Donnelly, and J. D. Woodhouse, *J. Vac. Sci. Technol.,* *B*[2]**1,** 1178 (1983).
115. K. Jain, C. G. Willson, and B. J. Lin, *Appl. Phys.* [*Part*] *B* **B28,** 206 (1982).

116. M. W. Geis, J. N. Randall, T. F. Deutsch, P. D. DeGraff, K. E. Krohn, and L. A. Stern, *Appl. Phys. Lett.* **43,** 74 (1983).
117. T. F. Deutsch and M. W. Geis, *J. Appl. Phys.* **54,** 7201 (1983).
118. P. W. Cook, S. E. Schuster, and R. J. von Gutfeld, *Appl. Phys. Lett.* **26,** 124 (1975).
119. D. J. Ehrlich, R. M. Osgood, Jr., and T. F. Deutsch, *Appl. Phys. Lett.* **38,** 399 (1981).
120. J. Y. Tsao and D. J. Ehrlich, *Appl. Phys. Lett.* **43,** 146 (1983).
121. D. J. Silversmith, D. J. Ehrlich, J. Y. Tsao, R. W. Mountain, and J. H. C. Sedlacek, *in* "Laser-Controlled Chemical Processing of Surfaces" (A. W. Johnson, D. J. Ehrlich, and H. R. Schlossberg, eds.), p. 55. Am. Elsevier, New York, 1984.
122. G. L. Loper and M. D. Tabat, *Appl. Phys. Lett.* **46,** 654 (1985).
123. G. L. Loper and M. D. Tabat, *Proc. SPIE—Int. Soc. Opt. Eng.* **459,** 121 (1984).
124. T. Donohue, *in* "Beam-Induced Chemical Processes, Extend Abstracts, 1985 Fall Meeting of the Materials Research Society" (R. J. von Gutfeld, J. E. Greene, and H. Schlossberg, eds.), p. 139. Materials Research Society, Pittsburgh, Pennsylvania, 1985.
125. R. T. Smith, J. D. Chlipala, J. F. M. Bindels, R. G. Nelson, F. H. Fischer, and T. F. Mantz, *IEEE J. Solid-State Circuits* **SC-16,** 506 (1981).
126. J. I. Raffel, A. H. Anderson, G. H. Chapman, K. H. Konkle, B. Mathur, A. M. Soares, and P. W. Wyatt, *IEEE J. Solid-State Circuits* **SC-20,** 399 (1985).
127. R. J. von Gutfeld and R. T. Hodgson, *Appl. Phys. Lett.* **40,** 352 (1982).
128. E. F. Krimmel, A. G. K. Lutsch, R. Swanepoel, and J. Brink, *Appl. Phys. [Part] A* **A38,** 109 (1985).
129. T. J. Chuang, *J. Chem. Phys.* **74,** 1461 (1981).
130. G. L. Loper and M. D. Tabat, *in* "Beam-Induced Chemical Processes, Extend Abstracts, 1985 Fall Meeting of the Materials Research Society" (R. J. von Gutfeld, J. E. Greene, and H. Schlossberg, eds.), p. 133. Materials Research Society, Pittsburgh, Pennsylvania, 1985.
131. J. P. Biberian and M. Ismeurt, *in* "Beam-Induced Chemical Processes, Extend Abstracts, 1985 Fall Meeting of the Materials Research Society" (R. J. von Gutfeld, J. E. Greene, and H. Schlossberg, eds.), p. 113. Materials Research Society, Pittsburgh, Pennsylvania, 1985.
132. T. J. Chuang, *J. Chem. Phys.* **73,** 6303 (1980).
133. G. Reksten, W. Holber, and R. M. Osgood, Jr., *J. Vac. Sci. Technol., A*[2]**2,** 50 (1984).
134. P. Brewer, W. Holber, G. Reksten, and R. M. Osgood, Jr., *Proc. SPIE—Int. Soc. Opt. Eng.* **459,** 128 (1984).
135. H. Okane, N. Hayasaka, S. Suto, M. Sekine, and Y. Horiike, *in* "Beam-Induced Chemical Processes, Extend Abstracts, 1985 Fall Meeting of the Materials Research Society" (R. J. von Gutfeld, J. E. Greene, and H. Schlossberg, eds.), p. 121. Materials Research Society, Pittsburgh, Pennsylvania, 1985.
136. M. R. Baklanov, I. M. Beterov, S. M. Repinskii, A. V. Rzhanov, V. P. Chebotaev, and N. I. Yurshina, *Sov. Phys.—Dokl. (Engl. Transl.)* **19,** 312 (1974).
137. I. M. Beterov, V. P. Chebotaev, N. I. Yurshina, and B. Ya. Yurshin, *Sov. J. Quantum Electron. (Engl. Transl.)* **8,** 1310 (1978).
138. A. W. Tucker and M. Birnbaum, *Proc. SPIE—Int. Soc. Opt. Eng.* **385,** 131 (1983).
139. N. Tsukada, S. Semura, H. Saito, S. Sugata, K. Asakawa, and Y. Mita, *J. Appl. Phys.* **55,** 3417 (1984).
140. M. Takai, J. Tokuda, H. Nakai, K. Gamo, and S. Namba, *in* "Laser-Controlled Chemical Processing of Surfaces" (A. W. Johnson, D. J. Ehrlich, and H. R. Schlossberg, eds.), p. 211. Am. Elsevier, New York, 1984.
141. P. Brewer, W. Holber, G. Reksten, and R. M. Osgood, Jr., *Proc. SPIE—Int. Soc. Opt. Eng.* **459,** 131 (1984).

142. P. Brewer, S. Halle, and R. M. Osgood, *in* "Laser-Controlled Chemical Processing of Surfaces" (A. W. Johnson, D. J. Ehrlich, and H. R. Schlossberg, eds.), p. 179. Am. Elsevier, New York, 1984.
143. D. J. Ehrlich, R. M. Osgood, Jr., and T. F. Deutsch, *Appl. Phys. Lett.* **36,** 698 (1980).
144. G. C. Tisone and A. W. Johnson, *in* "Laser Diagnostics and Photochemical Processing for Semiconductor Devices" (R. M. Osgood, S. R. J. Brueck, and H. R. Schlossberg, eds.), p. 73. Am. Elsevier, New York, 1983.
145. R. M. Osgood, Jr., A. Sanchez-Rubio, D. J. Ehrlich, and V. Daneu, *Appl. Phys. Lett.* **40,** 391 (1982).
146. R. D. Rauh and R. A. LeLievre, *J. Electrochem. Soc.* **132,** 2811 (1985).
147. P. D. Greene, *Proc. Int. Symp. GaAs Relat. Compd., 6th 1976,* p. 141 (1976).
148. R. P. Salathe and G. B. Rao, *in* "Laser Diagnostics and Photochemical Processing for Semiconductor Devices" (R. M. Osgood, S. R. J. Brueck, and H. R. Schlossberg, eds.), p. 65. Am. Elsevier, New York, 1983.
149. M. Takai, J. Tsuchimoto, H. Nakai, J. Tokuda, K. Gamo, and S. Namba, *in* "Beam-Induced Chemical Processes, Extend Abstracts, 1985 Fall Meeting of the Materials Research Society" (R. J. von Gutfeld, J. E. Greene, and H. Schlossberg, eds.), p. 129. Materials Research Society, Pittsburgh, Pennsylvania, 1985.
150. M. R. Aylett and J. Haigh, *in* "Beam-Induced Chemical Processes, Extend Abstracts, 1985 Fall Meeting of the Materials Research Society" (R. J. von Gutfeld, J. E. Greene, and H. Schlossberg, eds.), p. 63. Materials Research Society, Pittsburgh, Pennsylvania, 1985.
151. C. Uzan, R. Legros, Y. Marfaing, and R. Triboulet, *Appl. Phys. Lett.* **45,** 879 (1984).
152. R. Tenne, V. Marcu, and Y. Prior, *Appl. Phys.* [*Part*] *A* **A37,** 205 (1985).
153. V. Daneu, J. Peers, and A. Sanchez, *in* "Laser-Controlled Chemical Processing of Surfaces" (A. W. Johnson, D. J. Ehrlich, and H. R. Schlossberg, eds.), p. 133. Am. Elsevier, New York, 1984.
154. Z. E. Buachidze, I. V. Vasilishcheva, V. N. Morozov, V. A. Pletnev, A. S. Semenov, and P. V. Shapkin, *Sov. J. Quantum Electron.* (*Engl. Transl.*) **12,** 1514 (1982).
155. J. I. Steinfeld, T. G. Anderson, C. Reiser, D. R. Denison, L. D. Hartsough, and J. R. Holiahan, *J. Electrochem. Soc.* **127,** 514 (1980).
156. C. Fiori and R. A. B. Devine, *Appl. Phys. Lett.* **47,** 361 (1985).
157. J. M. Gee and P. J. Hargis, Jr., *Proc. SPIE—Int. Soc. Opt. Eng.* **459,** 132 (1984).
158. J. H. Brannon and T. J. Chuang, *in* "Beam-Induced Chemical Processes, Extend Abstracts, 1985 Fall Meeting of the Materials Research Society" (R. J. von Gutfeld, J. E. Greene, and H. Schlossberg, eds.), p. 147. Materials Research Society, Pittsburgh, Pennsylvania,1985.
159. S. Yokoyama, Y. Yamakage, and M. Hirose, *Appl. Phys. Lett.* **47,** 389 (1985).
160. B. T. Dai, B. S. Agrawalla, and S. D. Allen, *in* "Beam-Induced Chemical Processes, Extend Abstracts, 1985 Fall Meeting of the Materials Research Society" (R. J. von Gutfeld, J. E. Greene, and H. Schlossberg, eds.), p. 143. Materials Research Society, Pittsburgh, Pennsylvania,1985.
161. J. H. Brannon, *in* "Laser Chemical Processing of Semiconductor Devices, Extend Abstracts, 1984 Fall Meeting of the Materials Research Society" (F. A. Houle, T. F. Deutsch, and R. M. Osgood, Jr., eds.), p. 112. Materials Research Society, Pittsburgh, Pennsylvania, 1984.
162. A. W. Johnson and R. V. Smilgys, *in* "Laser Chemical Processing of Semiconductor Devices, Extend Abstracts, 1984 Fall Meeting of the Materials Research Society" (F. A. Houle, T. F. Deutsch, and R. M. Osgood, Jr., eds.), p. 108. Materials Research Society, Pittsburgh, Pennsylvania, 1984.
163. C. I. H. Ashby and P. J. Brannon, *Appl. Phys. Lett.,* **49,** 475 (1986).
164. G. L. Loper and M. D. Tabat, *J. Appl. Phys.* **58,** 3649 (1985).

Contacts to GaAs Devices

J. M. WOODALL, N. BRASLAU, AND J. L. FREEOUF

IBM Thomas. J. Watson Research Center
P. O. Box 218
Yorktown Heights, New York 10598

I. Introduction

One measure of the maturity of a device technology is the ease and reliability of applying contact metallurgy. For most successful and sophisticated metallurgies there usually exists a large and very diverse body of knowledge concerning the detailed behavior of the metal–semiconductor interface. Thus, the maturity of silicon device technologies is easily demonstrated by a perusal of those articles which deal with the metallization of silicon.

In contrast, the status of GaAs metallization is much different, and, to borrow an overworked pun, the surface has barely been scratched. Until recently, very little work had been done on metallizing GaAs integrated circuits. Instead, most GaAs devices such as lasers, solar cells, LEDs, and Gunn diodes were either discrete or monolithic devices for which the demands on metallization were rather modest. For example, acceptable contact resistivities for these applications are in the range 10^{-1}–10^{-5} Ω cm^2. However, with the advent of the GaAs MESFET and integrated circuits, very stringent requirements were placed on both ohmic and Schottky-barrier contacts. For source and drain ohmic contacts, a contact resistivity of less than 5×10^{-6} is generally required, while a Schottky-barrier height tolerance of ±0.01 eV is desired. As researchers worked to achieve these new goals, they discovered that basic information concerning both the metal–GaAs interface and the GaAs surface was lacking. As a result, during the past few years, there has been extensive research in the areas of: (1) alloyed dopant–metal contacts, particularly the Au–Ge–Ni contact, in order to lower contact resistances and improve uniformity; (2) the fundamental physics and chemistry of Schottky-barrier formation (including ohmic contacts); (3) new techniques for improving and controlling the properties of ohmic and Schottky contacts. This article will review the history and progress in the understanding of the metal–semiconductor interface and its application to the development of the ohmic contact for device fabrication.

II. "Ideal" Contacts

For an ideal metal–semiconductor interface, i.e., one in which the interface is inert and there are no appreciable surface or induced interface states in the semiconductor, the Schottky-barrier height is given by (*1*)

$$\phi_{bn} = \phi_m - \chi_{sc} \tag{1}$$

and

$$\phi_{bp} = E_g/q + \chi_{sc} - \phi_m \tag{2}$$

where ϕ_{bn} is the Schottky-barrier height to an n-type semiconductor (eV); ϕ_{bp}, the Schottky-barrier height to a p-type semiconductor (eV); ϕ_m, the metal work function (eV); χ_{sc}, the electron affinity of the semiconductor (eV); and E_g, the bandgap energy.

Thus, for example, an "ideal" ohmic metal–n-type-semiconductor contact is one in which $\phi_m \leq \chi_{sc}$ and $\phi_{bn} \leq 0$. Likewise, a rectifying Schottky-barrier contact is one in which $\phi_m > \chi_{sc}$ and $\phi_{bn} > 0$. Thus, for the ideal case and for a given semiconductor, ϕ_b should be determined by the metal work function. Unfortunately, this is not the case for GaAs and many other semiconductors.

III. The GaAs Surface and Interface—Fermi-Level Pinning

When a piece of GaAs is carefully cleaved in ultrahigh-vacuum (UHV) conditions, it is found that the position of the Fermi energy level at the (110) cleaved surface is usually the same as the Fermi level in the bulk (*2*), as shown in Fig. 1a. This has been found for other compound semiconductors as well (*2, 3*). It has also been shown that when 0.01–0.1 monolayer of different metals or oxygen cover the (110) surface of GaAs (and other compound semiconductors as well), the Fermi level at the surface becomes "loosely pinned" within a small range of energies, ±0.1 eV, and that the pinning energy is roughly independent of the "contaminating" metal or oxide (*4*). For GaAs this pinning position is about 0.8 eV below the conduction-band minimum (Fig. 1b). Since 0.8 eV is roughly the Schottky-barrier height most metals make to *n*-type GaAs (*5*) (and therefore independent of the metal work function), it has been postulated that Fermi-level pinning and Schottky barriers are determined by the same mechanisms.

Several models and empirical rules have been proposed to explain pinning. Surface states have long been invoked in discussions of Schottky barrier heights. Bardeen showed that a surface-state density of $\geq 10^{13}$

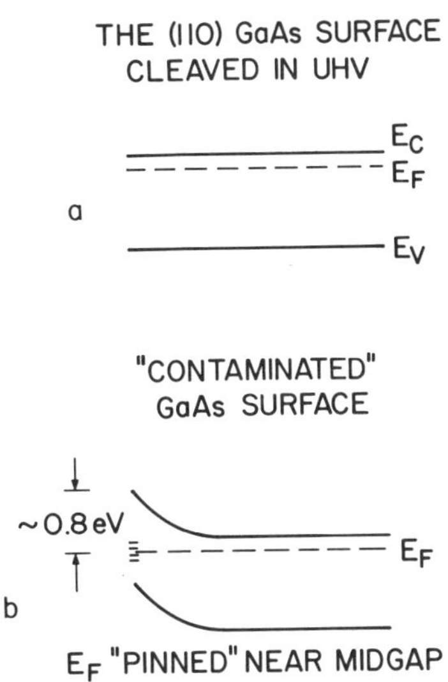

FIG. 1. Energy-band diagram of GaAs: (a) (110) surface cleaved in UHV conditions; (b) "contaminated" surface.

cm^{-2} is sufficient to fix the barrier height (*6*). Subsequently Mead and Spitzer (*5*) developed the two-thirds bandgap rule based on calculations of expected energies of the surface states. Kurtin *et al.* (*7*) subsequently showed that semiconductors which were more "ionic" in nature appeared more "ideal," i.e., had a lower surface-state density, hence $\phi_b = f(\phi_m)$, whereas the more covalent materials showed $\phi_b \neq f(\phi_m)$ behavior. Further studies by McCaldin *et al.* (*8*), using Au as a reference metal to a variety of *p*-type compound semiconductors, showed that ϕ_{bp} was related to the anion component of the semiconductor and that ϕ_{bp} varied inversely with the anion electronegativity. Since the valence-band maximum is due mainly to wave functions of the anion component of the semiconductor, it is suggested that the ϕ_{bp} is tied to the valence band, which is a bulk electronic property of the material rather than a surface property. Other approaches involve the concept of metal-induced gap states (MIGS), in which simple physical considerations of local charge neutrality suggest that near a metal–semiconductor interface the Fermi level in the semiconductor is pinned near an effective gap center (*9, 10*).

There are also models involving Fermi-level pinning which do not invoke surface states. Spicer *et al.* (*4*) have suggested that pinning is caused by native defects at the surface, e.g., vacancies and antisite defects, induced by the interaction of either metal or oxygen beams with the GaAs surface. Chemical reactivity of the metal–semiconductor interface has been shown by Brillson to be important in determining the variation in barrier height (*11*); he showed that semiconductors with relatively small heats of formations had barrier heights which were less sensitive to the applied metal than were semiconductors with relatively large heats of formation. The former type of semiconductor is expected to be very reactive with the metal contact, and thus the interface composition is expected to dominate the barrier height, whereas the latter type is less reactive, and thus the barrier height is expected to be determined by either the metal work function or by interface states induced by the metal. Zur *et al.* have recently quantified the effects of pinning of Schottky barriers (*12*). They have shown that for Schottky barriers formed by thick metal contacts, there can be only one pinning position for both *n*- and *p*-type materials and that regardless of the mechanism, pinning will not occur for "surface-defect" densities of less than about a monolayer.

In spite of the rich array of various models, there are some notable experimental results which remain unexplained. One is the fact that liquid gallium will make a temporary ohmic contact to lightly doped *n*-type GaAs under the conditions in which the native oxide to GaAs is disrupted, exposing clean gallium to an oxide-free GaAs surface (*13*). With time and exposure to air the contact will become rectifying, as predicted by previous models. Another troublesome observation is the difference in behavior between electroless Au and electroless Au–Sn contacts to lightly doped *n*-type GaAs. The Au contact is rectifying, whereas the Au–Sn contact is ohmic. Also, Pd–GaAs contacts exhibit unusual behavior. As deposited, Pd on GaAs is ohmic to *p*-type material and rectifying to *n*-type material. After an anneal at 500°C for 30 min, the contact is ohmic to both *n*- and *p*-type material. This is surprising since Pd is not known to dope GaAs. Another issue is the Okamoto *et al.* study (*14*) of Schottky-barrier heights for the Al–(GaAs–AlAs) interface prepared by molecular-beam epitaxy. They find barrier heights, particularly to AlAs, which are significantly different from those predicted by previous models and which are significantly different from those reported for Au–AlAs (*5*). Also puzzling is the case for Al contacts to $Ga_{0.5}In_{0.5}As$ deposited *in situ* at room temperature after the MBE growth of the GaInAs layer (*15*). These contacts are ohmic, whereas if the Al is evaporated after exposure of the GaInAs surface to air, a 0.2–0.3 eV Schottky barrier is formed. Finally, if Au is carefully deposited on UHV-cleaved GaAs (110) surfaces to a thickness of 10–20 monolayers, a barrier ϕ_{bn} = 1.1–1.3 eV is formed, which is significantly larger than those reported previously (*16*). It will be shown that these seemingly conflicting observations can be explained by a new model developed below. The new model also explains Schottky data for III–V materials previously reported in the literature.

The model is called the effective-work-function (EWF) model, in which the Fermi energy position at the surface (or interface) is not due to or fixed by surface states but rather is related to the work functions of microclusters of the one or more interface phases resulting from contamination prior to or reactions which occur during metallization. This behavior is shown in Fig. 2. The UHV-cleaved (110) surface is free of intrinsic surface states, and hence E_F is uniform, as seen in Fig. 1a. All other surfaces exhibit band bending prior to any intentional surface treatment or metal deposition. The theory requires that these "pinned" surfaces already contain microclusters of interface phases due to their exposure to air or any other surface-contaminating environment. When a metal is deposited, there is a region at the interface which contains a matrix of native oxide embedded with microclusters of different phases, each having its own work function. Since the model does not require "surface states," Eq. (1) can be modified and rewritten as

$$\phi_{bn} = \phi_{eff} - \chi_{sc} \tag{3}$$

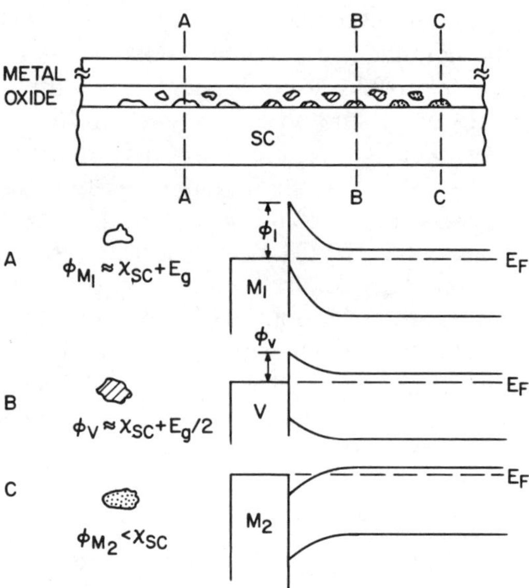

FIG. 2. Schematic diagram illustrating principles of the EWF model. ϕ_{M1}, ϕ_V, ϕ_{M2} are arbitrarily chosen to demonstrate the effect. ϕ_{eff} appropriately averages the effects of the various interface phases.

where ϕ_{eff} is an appropriately weighted average of the work functions of the different interface phases. Thus, the measured ϕ_{bp} can depend somewhat on the measurement technique, i.e., *C–V* or *I–V*. In other words, the interface phases comprise the Schottky-barrier contact. The rest of the bulk metallurgy has little or no effect on the barrier height, except when the interface phases are predominantly the same as the applied metallurgy.

For most of the III–V compounds, including GaAs, conventional metallization, i.e., non-UHV conditions, results in a condition in which ϕ_{eff} is due mainly to ϕ_V, to work function of the group-V component, and occurs as a result of either one or both of the following reactions

$$\text{V–O} + \text{III–V} \rightarrow \text{V} + \text{III–O} \tag{4}$$

$$M + \text{III–V} \rightarrow (\text{V}, \text{M}V_x) + (M, \text{III}) \tag{5}$$

where V–O and III–O are generic group-V and -III oxides and *M* is a metal.

The conditions for driving the reaction of Eq. (4) to the right and hence generating excess V at the interface is that the Gibbs free energy, ΔF, is negative. A list of such oxide reactions is shown in Table I (*17*). Note that

TABLE I

III–V OXIDE REACTIONS[a]

Material	ΔF
AlP	(−) Large
GaP[b]	(+) Small
InP[c]	(~) 0
AlAs	(−) Large
GaAs[d]	(−) Large
InAs[d]	(−) Large
AlSb	(−) Large
GaSb	(−) Large
InSb	(−) Small
$Ga_{1-x}Al_xAs$[d]	(−) Large

[a] E.g., $2GaAs + As_2O_3 \rightarrow Ga_2O_3 + 4As$
[b] Excess V not observed at interface.
[c] Excess P observed, 350–550°C anneal.
[d] Excess V observed at interface.

excess group V has been experimentally observed when ΔF is negative, i.e., GaAs, InAs, InSb (*18, 19*), and that it is not observed when ΔF is positive, i.e., for GaP (*19*). It is interesting to note that for InP, $\Delta F \sim 0$, it has been possible to form MISFET structures under special conditions which exhibit a low interface-state density (*20*). This is consistent with our model, which would predict either no or very little excess free phosphorus at the interface, since if it were there, phosphorus would form a positive Schottky barrier to InP, which would act like an interface state within the bandgap. Likewise, it should also be noted that for GaAs it is well known that MOSFET structures have notoriously high interface-state densities (10^{13}–10^{14} cm^{-2}) and that excess arsenic is usually observed at the interface (*21*). Again this is consistent with the model, since the ϕ_{bn} expected for the As–GaAs interface is about 0.8 eV (the barrier height usually observed for most metal depositions). Since workers have reported a large density of midgap states for MOSFET GaAs structures, the model would ascribe these "states" to arsenic clusters at the interface, which act as Schottky-barrier contacts with $\phi_{bn} \sim 0.8$ eV embedded in an oxide matrix. If the model is correct, a GaP MOSFET structure with low interface-state densities is predicted since no free P is expected at the interface. In addition to oxide reactions, excess group V can be generated by reaction of metals via the reaction of Eq. (5). For example, it is known that Au deposited on GaAs and GaP results in dissolved Ga in the Au film (*22*). Also, preliminary phase diagram data (*23*) show that an arsenic phase is expected at equilibrium for Au–GaAs and Au–InSb. Thus, a knowledge of both oxide and reactive metal chemistry should enable accurate predictions of the Schottky-barrier heights for metal–III–V interfaces.

A current status of the predictive ability of the model for III–V compounds is in Table II, which lists the experimentally derived values of ϕ_{bp} and $\chi_{sc} + \phi_{bn}$ for Au–III–V contacts, which were formed under conditions in which the group-V element is expected to be the dominant interface phase. There are three points to note in this table. First, Eq. (1), written as

$$5.1\text{–}5.5 \text{ eV} = \phi_{Au} = \chi_{sc} + \chi_{bn} \tag{6}$$

is not obeyed for the III–V compounds. Second, the common anion rule (*8*) is not obeyed for AlAs and AlSb. Third, the EWF model [Eq. (3)] agrees well, as expected, for $\phi_{eff} = \phi_V$. The Schottky-barrier results for the conditions in which the interface phase is predominantly that of the applied metallurgy is shown in Table III. It is seen that $\phi_m = \phi_{eff}$ for this case. For example, in the Al–AlAs case, this metallization was deposited and annealed in a UHV MBE system where the Al and the AlAs surfaces were very clean. Thus, it is expected that ϕ_{eff} should be dominated by $\phi_{Al} = 4.0\text{–}4.2$ eV. Thus, since $\chi_{sc} + \phi_{bn} = 4.2\text{–}4.6$ eV for this case is much closer to ϕ_{Al} than to ϕ_{As} or ϕ_{Au}, we believe that ϕ_{bn} is due mainly to ϕ_{Al}. Similarly the Ga–GaAs ohmic contact mentioned earlier can be ex-

TABLE II

Au–III–V SCHOTTKY BARRIERS[a]

Anion	III–V	ϕ_{bp}	$\chi_{sc} + \phi_{bn}$	ϕ_V
P	GaP	0.96	4.9	5.0[c]
	InP	0.85	4.9	5.0[c]
As	AlAs[b]	0.9	4.7–5.1	5.0[c] (4.8)[M]
	GaAs	0.5	5.0	5.0[c]
	InAs	0.3–0.5	4.8–5.0	5.0[c]
	GaInAs	0.7–0.5	4.8–4.9	5.0[c]
Sb	AlSb[b]	0.54	4.7	4.8[c] (4.7)[M]
	GaSb	0.1	4.7	4.8[c]
	InSb	~0.1	4.8 (77 K)	4.8[c]

[a] For cases in which excess group V is expected at interface. Question: Does $\chi_{sc} + \phi_{bp} = \phi_{Au} = 5.1\text{–}5.5$ eV? Or is $\chi_{sc} + \phi_{bp} = \phi_V$ a better match? All values from citations in Freeouf *et al.* (*17*). M are measured values.

[b] Does not obey common anion rule.

[c] Theoretical prediction.

TABLE III

SPECIAL CASES FOR EWF MODEL[a]

Contact	$\chi_{sc} + \phi_{bn}$ (eV)	ϕ_m (eV)
Au/GaAs [UHV (110)]	5.2–5.4	5.1–5.5
Ga/GaAs (oxide disrupted)	4.1–4.4	4.3–4.4
Au–Sn/GaAs (electroless, Sn dominant on *n*-type material, I–V)	4.1–4.4	4.4
Al/AlAs (MBE/anneal)	4.2–4.6	4.2–4.3
Al/$Ga_{0.5}In_{0.5}As$ (MBE/room temp.)	≤4.4	4.2–4.3

[a] Does $\chi_{sc} + \phi_{bp} = \phi_m$, the work function of the applied metal, as is expected from the experimental conditions?

plained since $\phi_{Ga} = 4.36$ eV (*24*) and $\phi_{bn} = 0$–0.3 eV (for ohmic behavior); $\chi_{GaAs} + \phi_{bn} = 4.1$–$4.4$ eV $\sim \phi_{Ga}$.

Next, we tentatively suggest that the EWF model may be able to explain the results on Fermi-level pinning experiments performed in UHV conditions on cleaved surfaces of III–V materials, e.g., GaAs, particularly those of Spicer *et al.* (*4*) and Skeath *et al.* (*16*). They observe a more or less well-defined pinning energy ~0.8 eV above the valence band for *n*-GaAs (110) for overlayer coverages of 01.–1.0 monolayers and that the pinning energy is independent of material (except for Au coverage) for materials such as Cs, Al, Ga, In, and O. They interpret this result to mean that the pinning must be due to native defects, i.e., Ga and/or As vacancies in the surface of the GaAs, which are generated by impingement or adsorption of the coverage material; thus the pinning energy corresponds to the energy of the defect. We note that the "pinning" position of 0.8 eV is not incompatible with the ideas of the EWF model if the impinging atom were either oxygen or atoms with $\phi \sim \phi_{Ga}$, and it is assumed that the impinging atom knocks out both a Ga and an As atom pairwise from the lattice onto the surface. For this case, ϕ_{eff} would be an appropriately weighted average of the work functions of the atomic species in the surface.

$$\phi_{eff} = \frac{W_{dGa}\phi_{Ga} + W_{dAs}\phi_{As} + W_{dm}\phi_m}{\Sigma W_d} \tag{7}$$

For the assumptions cited above, an impingement of 1 Ga atom onto a GaAs surface would result in 2 Ga "adatoms" and 1 As "adatom." Since the density of states for Ga is roughly one-half (*25*) that of As, a possible averaging scheme (ignoring screening by the substrate) would be

$$\phi_{eff} = (2\phi_{As} + 2\phi_{Ga})/(\sim 4\text{–}4.5 \text{ eV}) \quad (8)$$

Therefore, $\phi_{bn} \sim 4.5\text{–}4.0 \sim 0.5$ eV, and Ef ~ 0.9 eV from the valence band, as reported in Spicer *et al.* (*4*). UPS should observe such an averaged band bending if the Ga and As atoms do not cluster into well-defined separate regions of lateral dimensions comparable to or larger than the Debye length of the substrate [~3 nm Spicer *et al.* for (*4*)]. A cluster size larger than this value would result in UPS observations of two well-defined but different band-bending values, whereas cluster sizes less than this should result in an averaging similar to that discussed above (*26*). Thus, apparent pinning may in fact merely be a "Schottky barrier" determined by the "average" work function of the adatoms or microclusters. Since for *n*-type GaAs with doping of 10^{17}–10^{18} cm^{-3} the necessary surface charge to create a band bending of 0.6–0.8 eV is only $\sim 10^{12}$ cm^{-2}, 0.1 monolayers ($\sim 10^{13}$ atoms cm^{-2}) is more than sufficient to absorb the charge. If the 0.1 monolayer is composed of either single atoms or clusters, charge transfer between the various isolated adatoms and clusters can occur via the large Debye lengths in the semiconductor (>3 nm).

For the case of Au on GaAs (*16*), and coverages of 0.1–10 monolayers, the measured surface Fermi energy varies continuously from the previously reported 0.8 eV position for 0.1 monolayer to ~0.3 eV at 10 monolayer coverage. This result is hard to explain by the adatom-induced-defect model; however, the EWF model suggests that, as the Au coverage increases, Au becomes the "dominant" species, and $\phi_{Au} = 5.1\text{–}5.2$ eV $\sim \phi_{eff}$. Hence, $\phi_{bn} \sim 1.1$ eV would be the expected result of Skeath *et al.* (*16*). Furthermore, at large coverages, the pinning position acts as one expects Schottky barriers to act; i.e., the pinning position is independent of bulk doping.

Finally, we ask the question: If there are $\sim 10^{12}$ defects cm^{-2} ($>10^{19}$ cm^{-3}) which pin the Fermi-level midgap, how can workers grow GaAs with slightly compensated carrier concentrations of only 10^{14} cm^{-3} by MBE? By what mechanism do the $\sim 10^{12}$ defects cm^{-2} get reduced to $\sim 10^7$ during growth? We note again that, since the EWF model does not invoke defects *within the semiconductor* it allows for the straightforward growth of low-defect-density crystal layers via vapor deposition as in MBE.

We expect, then, that the GaAs surface under the conditions found in device processing will be pinned at 0.8 V, with a depletion region beneath the surface whose thickness depends in the usual way of the doping density. It is the presence of this depletion region which is the source of the difficulty in obtaining ohmic contacts.

IV. Alloyed Ohmic Contacts to GaAs

1. Introduction

With the explosive growth of GaAs as a useful semiconductor for microwave, digital, and optical devices, much attention has been devoted to the development of ohmic contacts, through which we have to communicate with the interior of the device from the outside world. Ideally, we desire an interface to an external metal which is linear, noninjecting, smooth, spatially abrupt, stable in time with impressed voltage, with a small resistance to current flow, and, hopefully, one that does not require elaborate processing steps. These properties are required over a four-decade range of doping concentrations, from very lightly doped Gunn oscillators ($N_D \sim 10^{14}\ cm^{-3}$) to degenerately doped injection lasers ($N_D \sim 10^{18}\ cm^{-3}$).

The alloyed AuGe-based contact is widely used to make ohmic connections to GaAs (*27*). It has been presumed that the regrown alloyed region is heavily doped so that carrier transport is by tunneling through a very thin depletion region. The electrical and metallurgical properties of this heterogeneous system have been extensively studied and have been shown to be spatially nonuniform. Details of fabrication technique, analysis, and theoretical interpretation of its behavior will be discussed. It is suggested that the observed inverse doping dependence of the contact resistivity is due to spreading resistance domination of current paths through submicrometer regions of the contact area where heavy doping occurs (*28*).

As previously discussed, under ordinary processing conditions the Fermi level is pinned near midgap so that deposition of most metals onto a cleaned *n*-GaAs surface makes a Schottky barrier with a barrier height ~0.8 eV. For the range of doping concentrations employed in device fabrication, electron transport through such a barrier is by thermionic emission, and the current–voltage characteristic is diodelike, $J = J^s(\exp qV/kT - 1)$, where the parameter appropriate to our discussion, the contact resistivity, is (*29*)

$$r_c = \left.\frac{\partial J}{\partial V}\right|_{V=0}^{-1} = \frac{kT}{qJ_s} \sim 10^6 \quad \Omega\ cm^2 \tag{9}$$

and J_s is independent of doping.

In order to lower this contact resistivity, it is necessary either to reduce the barrier height (which is not easy) or to increase the doping at the metal–semiconductor interface to a high value so that the current transport through the thinner depletion layers is enhanced by tunneling. Then the voltage drop across the contact is of the order of kT/q, the diode equation will be linear to first order, and the contact is ohmic, with $r_c < 10^{-3}\ \Omega\ cm^2$ for n-type material.

The earliest contacts to GaAs were made with Sn, alloyed at temperatures of the order of 450°C (*30, 31*). However, this contact tended to be nonuniform on the surface, and the tin proved to be a very fast diffuser under applied field, tending to form conducting channels. Braslau *et al.* (*32*) introduced an evaporated AuGe eutectic with a Ni overlayer, alloyed at a temperature greater than the eutectic melting temperature, which overcame these problems and has become a very widely used technique for contacting n-GaAs. Other evaporated and alloyed systems such as AgInGe (*33*), AuGe/AgAu (*34*), and In/$AuGe_3$ (*35*) are also used. In all these systems, Ge is shown to be a donor in GaAs when on a Ga site, and Ga vacancies are produced in the alloying process, which are populated by indiffusing Ge to densities of the order of $10^{19}\ cm^{-3}$. The surface morphology is not uniform, and there is evidence of much structure as a function of depth. The extensive use of this contact has generated a large body of literature in which the electrical and metallurgical properties have been investigated in some detail (*36*).

One interesting feature is an apparent N_D^{-1} dependence of r_c (Fig. 3). From tunneling theory one would expect an $\exp(1/\sqrt{N_D})$ dependence for large doping concentration N_D (*29*), but it is difficult to explain a subtle variation of the Ge doping as a function of the underlying material doping to give an N_D^{-1} behavior.

Experience has shown (*36, 37*) and theory leads one to expect (*29*) that it is not as difficult to contact p-GaAs. In addition, most p-type layers of device interest are heavily doped. This aspect of contact technology will not be discussed further here, except to point out that heterojunction bipolar technology is strongly limited by contact resistance to the p^+ base region using currently available alloyed contacts (*38*).

2. Fabrication

By the term alloying, we imply a heating of the sample above the evaporated film's melting temperature and a subsequent cooling and regrowth of some heterogeneous structure. Alternately, sintering is performed at temperatures below the eutectic melting point, resulting in solid-phase epitaxy (*35*). The Au–Ge eutectic melts at 360°C. In either

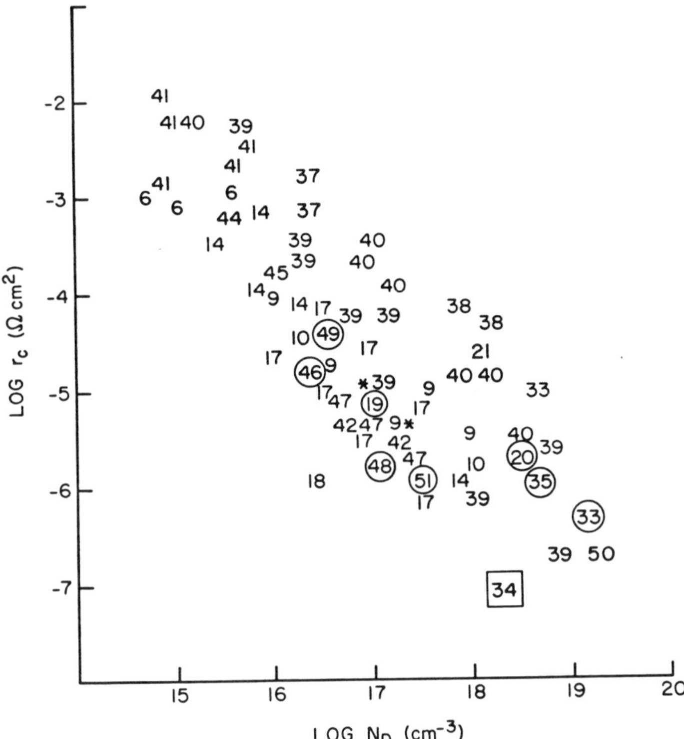

FIG. 3. Observed contact resistivity as a function of *n*-GaAs doping. The numbers identify reference number in Braslau (*28*) of the cited point. Circled points are laser or electron-beam alloyed. The point in the square is for an MBE-grown heterojunction. [From Braslau (*28*).]

case, GaAs dissolves until one component reaches saturation in the melted metal layer.

The process parameters available are:

(1) Surface preparation prior to deposition,
(2) Order and thickness of selected metal depositions,
(3) Use of a capping layer, and
(4) Alloying temperature–time profile.

Typically, samples are rinsed in organic solvents, etched lightly, rinsed in DI water, dried carefully, and placed in the vacuum chamber. We have found that the water rinse is necessary to remove hydrocarbon residues from the solvents. In any case, a thin oxide layer will form before the chamber can be evacuated. Some workers heat the surface to desorb the oxides (*39*), and others sputter clean the surface (*40*), although sputtering may introduce damage centers which give poor long-term integrity. Variations in the state of the surface at the beginning of the process probably contributes to the difficulty in resolving differences in details of the electrical and material properties of the resulting contacts reported by various workers. The thin oxide layer will be there and electrically is not significant since it is easily tunneled through, but the interface is certainly modified and influences the subsequent surface reactions.

Evaporation is usually done with the substrate at room temperature. If liftoff is used, it is necessary to hold the sample below 100°C to prevent polymerization of the photoresist. Au–Ge is evaporated from the eutectic (88% Au, 12% Ge by weight). Because of the differences in vapor pressure, we usually evaporate to completion a measured amount using a resistance heater. It was originally thought (*32*) that the subsequently deposited Ni layer served to hold down the AuGe and prevent it from balling, but it is now recognized that the Ni improves the wetting of the eutectic film and increases the solubility of GaAs. Sometimes a thin Ni film is deposited first to enhance wetting, but whether the Ni is on top or bottom probably does not make much difference when the film melts.

Typical thicknesses are 1000–1500 Å for AuGe and 100–500 Å for Ni. In spite of its usefulness, Ni is a very fast diffuser in GaAs, and excessive amounts degrade contact performance (*41, 42*). Additional Au is sometimes deposited to reduce the sheet resistance of the contact metal, but excess Au also degrades contact performance (*43*) as it leads to excessive Ga outdiffusion, leaving excess Ga vacancies with not enough Ge to fill these donor sites. This can be controlled by placing a refractory barrier film between the AuGe/Ni and the thicker Au layer (*43*). However, thin layers give better aging characteristics (*44*).

Vidimari (*45*) reported improved contacts using a SiO_2 film some 100 Å thick in place of a Ni overlayer (but with a thin Ni film under the AuGe). This film is chemically inert during alloying but exerts a stress tending to hold the film on the surface. This process has been used in our laboratory (*46*); we see evidence of smaller rms roughness of the alloyed film, which is useful, for example, if it must serve as the bottom plate of a planar rf capacitor.

Reported optimum temperature–time profiles vary widely. We use a horizontal furnace in which the sample is rapidly (400°C/min) brought to 450°C, held for 2 min, then rapidly cooled in an atmosphere of hydrogen or forming gas. There is evidence that rapid heating and cooling give better results (*37, 47*). Temperatures of 400°C have been reported to give better long-term behavior (*44*), since the reaction has been found to produce a less spatially unstable interface. Each laboratory seems to have its own favorite formula, and none has been shown to be much superior to another.

A promising alternative to the classical alloying process described above is to use laser or electron beams. A recent review of this work can be found in Eckhardt (*48*). Superior surface morphology is claimed for this technique, as well as decreased interdiffusion and segregation because of the short heating times involved. The best contact resistivity reported for those AuGe-based systems is comparable to that obtained by furnace annealing. Laser alloying of an evaporated Ge film with a Au layer later electroplated has also been reported (*49*).

3. Characterization

The magnitude of the electrical parameter of the contact, r_c, can be found only by evaluating the resistance of a metal–contact–semiconductor–contact–metal sandwich and subtracting out the unwanted contributions. Contact resistance depends on geometrical effects and current crowding, as well as on the contact resistivity of the interface. For bulk-grown thick wafers, the method of Cox and Strack (*33*) is used, where a series of contact pads of different radii are made on the top surface of the wafer and the resistance is measured between one of them and a large area contact on the bottom surface (see Fig. 4a). This resistance is a series combination of a fixed large-area contact resistance, a spreading resistance proportional to ρ/d, where ρ is the semiconductor resistivity and d

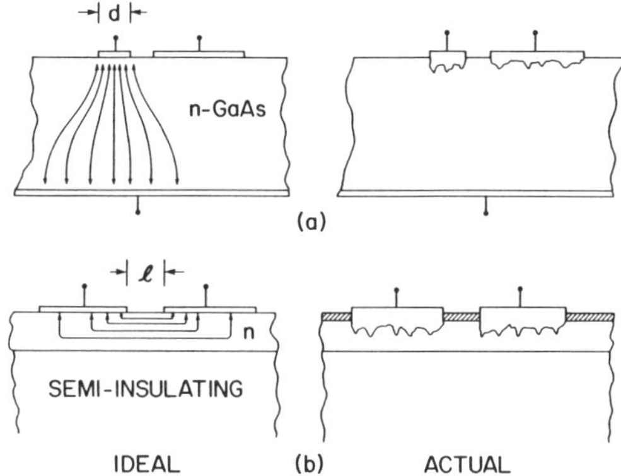

FIG. 4. Idealized and realistic cross sections of device geometries used to measure contact resistivities; (a) the method of Cox and Strack (*33*), in which current lines are normal to the interface; (b) the TLM method (*50*), in which the current lines are assumed parallel to the surface. [From Braslau (*28*).]

the contact diameter, and the contact resistance, $4r_c/\pi d^2$, where the current lines are assumed normal to the contact interface. For small r_c, small-diameter contacts are needed to avoid large errors in its determination. For planar devices, the current flow is predominantly parallel to the surface, and r_c is usually measured by the transmission-line method (*50*) (Fig. 4b). The contact is assumed to be an equipotential surface of zero thickness, and the conducting channel under it is treated as a resistive line whose series elements are proportional to the sheet resistance of the channel and whose shunt elements are proportional to r_c. The measured resistance between two square pads of width z separated by a distance l on a channel of sheet resistance $R_{\square}$ is given by

$$R = \frac{R_{\square}}{z}\left(l + 2l_T \coth \frac{z}{l_T}\right), \qquad \text{where} \quad l_T = \left(\frac{r_c}{R_{\square}}\right)^{1/2} \tag{10}$$

A linear plot results for the resistance as a function of separation; the slope yields $R_{\square}$, and the intercept with the abscissa yields twice the contact resistance, enabling r_c to be found. If r_c is small, $R_{\square}$ must be small to give a reasonably accurate value. If the channel is thick, correction must be made for current flow perpendicular to the surface under the contact. If the contact metallurgy alters the underlying channel sheet resistance, as illustrated in Fig 4, the actual value of r_c cannot be found from this one measurement. The value of R at $l = 0$, however, is unambiguous and gives twice the resistance of a contact of width z, so that source resistance per unit MESFET width can be found this way.

Procter *et al.* (*51*) have described a Kelvin resistor structure from which r_c can be directly determined. When the resistance of the contact is small with respect to the sheet resistance, as is found for GaAs samples, lithographic control and alignment are very critical. To our knowledge, no measurements by this technique have yet been reported for GaAs.

Microspot Auger electron spectroscopy sputtering profiles have proven to be very useful in interpreting the complex nature of the contact interface. Together with scanning electron microscopy, Rutherford backscattering, optical microscopy, and x-ray diffraction, a detailed picture of the contact region has been obtained. One of the most important findings is that the alloy is as much as 1900 Å deep (*41*), which may be a significant fraction of the depth of the conducting channel in a planar device. The Ge and Ni can penetrate deeply but uniformly into the GaAs, leaving microscopic grains of Ge-rich material, probably Ni–As–Ge, with Au–Ga grains also formed. The presence of Ni enhances the interdiffusion of Ge and GaAs. Variations in film thickness and alloying time profiles modify the degree of the roughness of the interface, but not its basic nature (*34*).

4. Models of the Contact

If Ge does provide a heavily doped layer at the metal–semiconductor interface, we expect a band model as shown in Fig. 5. If the contact is planar and uniform, the Ge layer must be heavily doped (*29*), $n^+ > 5 \times 10^{19}\ \mathrm{cm}^{-3}$, to give $r_c \sim 10^{-6}\ \Omega\ \mathrm{cm}^2$, as has been reported. The ideal n–n^+ interface should not by itself add any series resistance to the tunneling barrier if the majority carriers are thermalized. Popovic (*52*) has discussed transport through such a band structure where the n^+ region is thin compared to the carrier mean free path and derives a contact resistivity proportional to N_D^{-1}, the doping in the underlying region, since only those ballistic electrons with energies exceeding the n–n^+ barrier height can contribute to the current. There is too much evidence, however, that this doped region is at least 1000–2000 Å thick (*41*), which is more than adequate for carrier thermalization. The series resistance of such a heavily doped layer is negligible. There is also much evidence that the actual contact is nonplanar and nonuniform. Figure 6 is an optical microphotograph of a section of a TLM sample where the contact has been removed. A well-defined random array of pits can be seen.

We have proposed (*28*) a contact model which explains the observed contact resistivity and, in particular, the observed N_D^{-1} dependence. We assume the current flows through the Ge-rich islands, which are connected together through the overlying metal, as shown in Fig. 7 [See, for

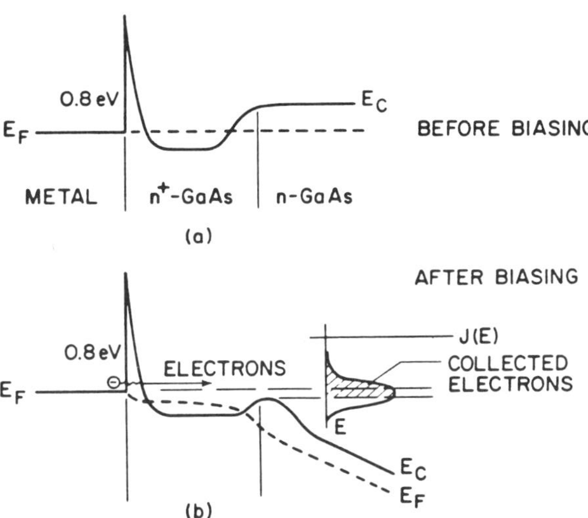

Fig. 5. Band model of metal–GaAs contact before and after bias is applied. [From Braslau (*28*).]

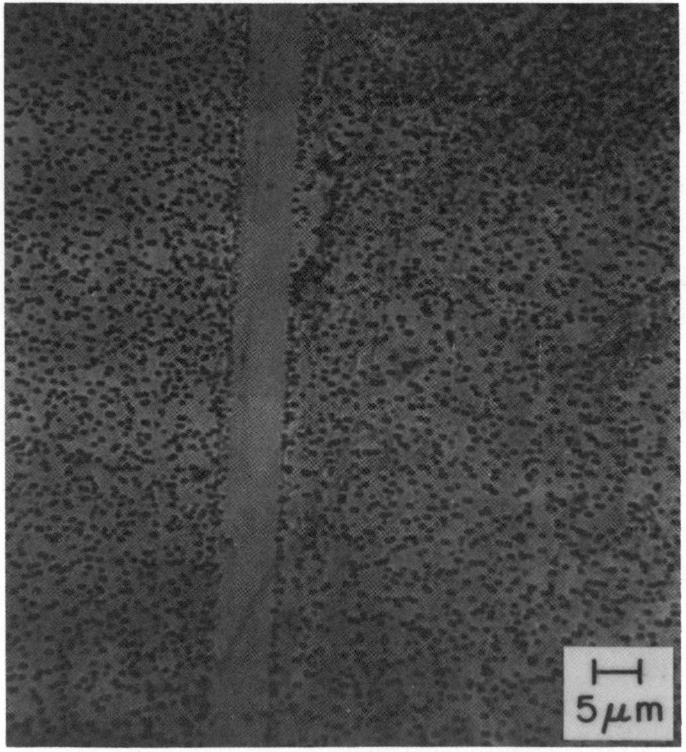

FIG. 6. Optical microphotograph of alloyed AuGeNi contact after contact metal has been removed, showing irregular pitting of GaAs surface. [From Braslau (*28*).]

example, Fig. 10 of Miller (*34*).] The real contact area is the approximately hemispherical region of radius r whose true contact resistivity r'_c is reduced from that expected, due to field enhancement (*53*) at these penetrating points. In the intermediate Ge-poor regions the conduction is much smaller due to the exponential dependence of tunneling current on underlying doping and, in any case, are shorted by the conducting protrusions. Assuming these regions are connected by the metal layer, one would measure a contact resistivity

$$r_c = |a|^2\left(\frac{\rho}{\pi|r|} + \frac{r'_c}{2f|r|^2}\right) \tag{11}$$

where $|a|$ is the mean separation of the protrusions and $|r|$ their mean radius; ρ is the resistivity of the region of doping N_D; and the field enhancement factor (*53*) $f >> 1$. For $\rho > 10^{-3}\ \Omega$ cm, the second term may

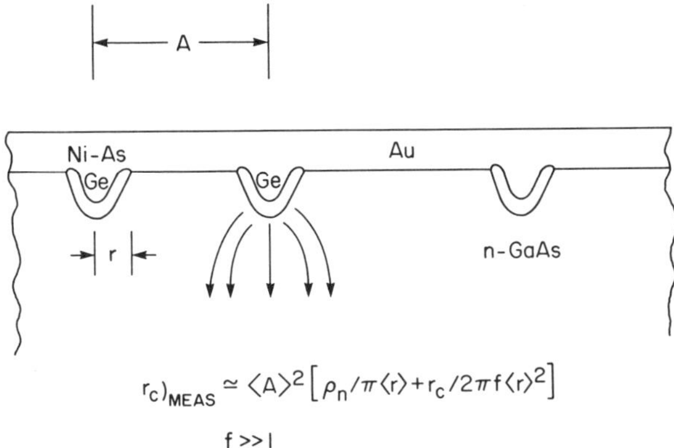

FIG. 7. Suggested model of alloyed ohmic contact to GaAs. Conductance is through a parallel array of Ge-rich protrusions of negligible contact resistance compared to spreading resistance in series with them. [From Braslau (*28*).]

be neglected. The contact resistivity is proportional to N_D^{-1} since ρ is $(N_D q\mu)^{-1}$, with μ the mobility and q the electronic charge.

If we estimate $|a|$ and $|r|$ from Fig. 6 or from Fig. 15b of Miller (*34*), Eq. (11) gives the solid line superimposed on the data of Fig. 3, shown in Fig. 8. Thus the contact resistivity is really dominated by spreading resistance in most alloyed contacts. This picture is supported by high-resolution TEM and STEM studies on this alloyed contact by Kuan (*54*). Even beam-alloyed contacts show this trend, although their top surfaces are smoother than furnace-alloyed contacts. It is likely that the semiconductor–contact interface is similar for both processes.

There are plenty of opportunities for additional pathology. Excessive Ga outdiffusion may leave Ga vacancies or Ni may act as a *p*-type dopant and compensate the Ge (*46*). Magee *et al.* (*55*) have found evidence of Cr redistribution during annealing into regions of near surface damage because of strain due to deposition of Au layers. Since many of the wafers which conducting layers have been grown on or implanted into are Cr-doped semi-insulating substrates, such redistribution is expected to further complicate the metallurgy described above.

If the size and spacing of these protruding contacting regions are difficult to influence in processing, then the only practical way to decrease contact resistance is to alloy to a heavily doped region previously constructed by other means (multiple ion implants, molecular-beam epitaxy, etc.), thus reducing the spreading resistance before the *n*-doped channel is reached. This has been shown to give reduced source resistance in FETs (*56, 57*).

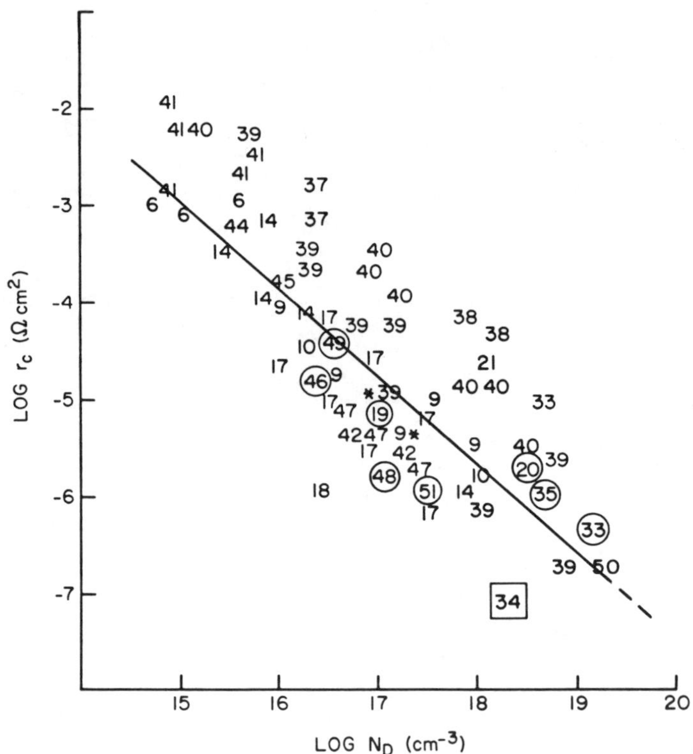

FIG. 8. Data shown in Fig. 3 with contact resistivity derived from Eq. (11), using mean values of a and r from Fig. 6. [From Braslau (*28*).]

5. ALTERNATIVE CONTACT METHODS

If the heavily doped layer can be grown with sufficient numbers of donors, it is possible to form a tunneling contact simply by depositing a metal on the surface (*58–61*). The resulting contact resistivity agrees well with the tunneling prediction for $n^+ > 10^{19}$ cm^{-3} (as measured from sheet resistance) (*61*). Similar results were obtained (*62*) by depositing refractory metals on electron-beam-annealed Se^+-implanted GaAs, which provided $n^+ > 10^{19}$ cm^{-3}. The ability of n^+-dope GaAs by laser-assisted diffusion of Sn from spun-on SnO_2 has been reported (*63*).

As the tunneling probability over a barrier depends exponentially on its height, much can be gained by lowering the metal–GaAs barrier height. One way is to grow by MBE a heavily doped Ge layer on top of a heavily doped GaAs layer, also MBE grown. Deposition of metal on Ge results in a Schottky barrier height several hundred millivolts below that found on

GaAs, and the series combination of this reduced barrier and the n^+-Ge–n^+-GaAs heterojunction gives a smaller contact resistivity than can be found from depositing directly on GaAs (*64*).

A possible technique for directly lowering the barrier height has been reported (*65*), whereby the Fermi-level pinning is modified by adsorption of H_2S. This gives an apparent barrier height of 0.4 eV; subsequent *in situ* deposition of Al films provides an ohmic contact consistent with that barrier height.

V. Heterojunction Contacts

Fortunately, there are ways to circumvent the problem of uncontrolled Schottky barrier heights. One of the most widely used methods is the lattice-matched heterojunction. When two semiconductor materials with nearly equal lattice constants form an epitaxial interface, predictable interface transport properties are observed, provided very abrupt junctions can be formed and one material does not dope the other. An example is the $Ga_{1-x}Al_xAs$–GaAs interface shown schematically in Fig. 9. When the electron affinities are different, $\chi_A \neq \chi_B$, a rectifying Schottky barrier with a controlled barrier height is formed (Fig. 9). When $\chi_A \sim \chi_B$, ohmic behavior is observed. This situation is thought to occur for the n-Ge–n-GaAs interface (*64*); however, there is some chance that the ohmic behavior could be due to degenerate doping of the GaAs side of the interface by Ge.

The Fermi level does not always pin in midgap. For example, surfaces of InAs exhibit pinning in the conduction band (*66–68*), as predicted by the EWF model. Thus, the situation for a metal–n-InAs contact shown in Fig. 10 produces an "ideal" ohmic contact where ϕ_b is ≤ 0. In this case, tunneling is not required, and low-resistance contacts can be made for a

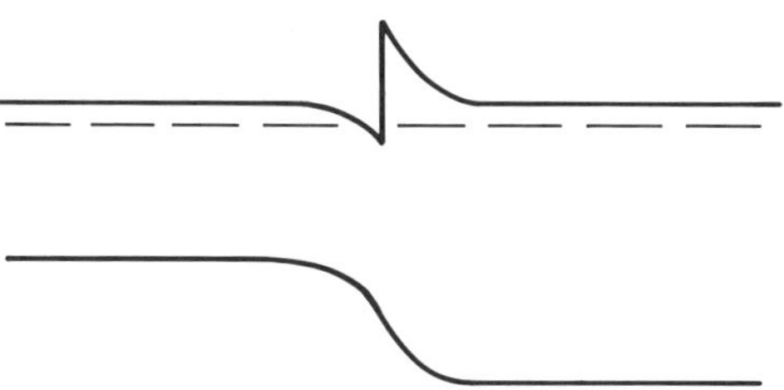

FIG. 9. Expected band-alignment diagram for a lattice-matched heterojunction, e.g., GaAs–GaAlAs, where $\chi_A \neq \chi_B$. [From Woodall *et al.* (*72*).]

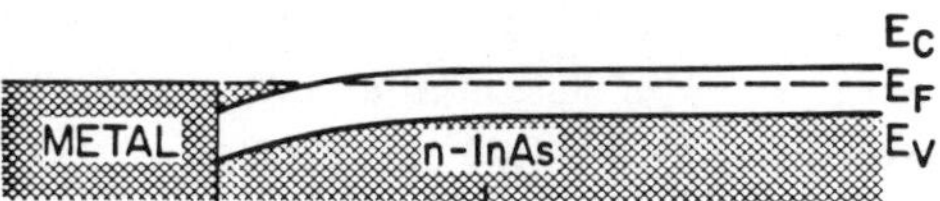

FIG. 10. Band diagram for M–n-InAs contact. [From Woodall *et al.* (*72*).]

wide range of n-type doping without need of alloying to form n^+ surface layers. With this in mind, one might conclude that good ohmic contacts for GaAs would result if the structure M–n-InAs–n-GaAs were used. However, this is not the case, and the reason for this is shown in Fig. 11. For this structure, there is a positive ϕ_b between the n-InAs and n-GaAs, which, depending on the doping level, results in either rectifying or tunneling ohmic contacts. This barrier results from one or more of the following: (1) a large electron-affinity discontinuity across the interface, (2) a large lattice-constant discontinuity leading to misfit dislocations, and (3) a "dirty" GaAs surface prior to epitaxial growth. The effect of the latter two is to produce midgap interface states and hence midgap Fermi-level pinning (*69, 70*). Thus, the n-InAs–n-GaAs abrupt junction behaves like the M–n-GaAs contact. A solution to this problem is shown in Fig. 12. For this case, the abrupt n-InAs–n-GaAs junction is replaced by a layer of $Ga_{1-x}In_xAs$ graded in composition from $x = 0$ at the GaAs interface to $x = 1$ at the InAs interface. [Alternatively, the InAs can be omitted entirely and the metal deposited directly on the graded layer. For this case, the surface of the layer can have a composition (*71*) $0.8 \leq x \leq 1$]. Notice that for Fig. 12 there are no abrupt discontinuities in the conduction band and that ϕ_b is ≤ 0 for the M–n-InAs contact. Thus this structure is expected to produce nonalloyed low-resistance ohmic contacts (*72*).

This structure was fabricated by MBE. It consisted of a 0.5 μm layer of Ge-doped n-GaAs ($\sim 2 \times 10^{17}/cm^3$) deposited directly onto the substrate, followed by a graded layer of 0.25 μm thickness, starting at $Ga_{0.999}In_{0.001}As$ and ending with InAs. The donor concentration was

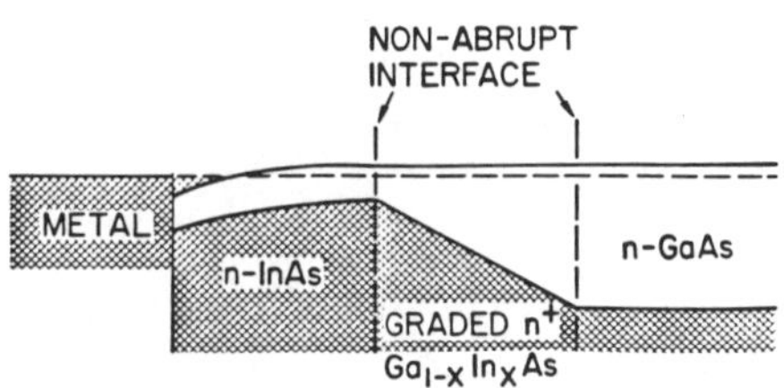

FIG. 11. Band diagram for M–n-InAs–n-GaAs structure. [From Woodall *et al.* (*72*).]

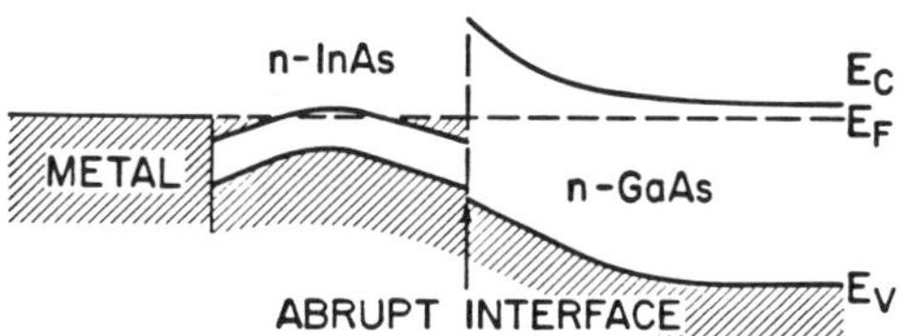

FIG. 12. Band diagram for ohmic-contact structure: M–n-InAs–n-GaInAs–n-GaAs. Note absence of barriers to electron flow in this structure. [From Woodall *et al.* (*72*).]

raised to $\sim 3 \times 10^{18}$ cm^{-3} in the first 0.1 μm of the graded layer. A layer of Ag (0.1 μm) was deposited *in situ* after the growth of the last layer.

The contact resistivity was determined from a transmission-line-type measurement to be $5 \times 10^{-7} < r_c < 1 \times 10^{-6}$ Ω cm^2. The present results are entirely consistent with the energy diagram shown in Fig. 12. Furthermore, Kajiyama *et al.* (*71*) have reported the compositional dependence of ϕ_b upon In concentration in $Ga_{1-x}In_xAs$, and it appears likely that this technique can be used to obtain a Schottky barrier height anywhere between that of GaAs ($\sim$0.8 eV) and that of InAs (<0 eV) with the same metal.

VI. Recent Results

As discussed earlier, the major reason that Fermi-level pinning at 0.8 eV below the conduction band is detrimental to ohmic-contact technology is that, in order to obtain tunneling contact resistivities as low as 10^{-6} cm^2 (a conservative estimate for high-performance devices), the space-charge density at the contact interface needs to be as high as 10^{20} cm^{-3}. From studies of n-type doping of GaAs it has been found that the maximum electron concentration (hence, the donor minus acceptor density) is about 1×10^{19} cm^{-3}. This concentration leads to a contact resistivity of about 100×10^{-6} Ω cm^2 and is much too large for both microwave and digital applications. Except for the notable exceptions of Refs. *59* and *62*, further increase in donor density does not increase the electron concentration. This has led to a belief among workers in this field that attempts to reduce contact resistance via higher doping will be unsuccessful. However, Kirchner *et al.* (*73*) recently reported studies of MBE growth of GaAs doped with Si where nonalloyed contact to GaAs with a Si concentration of 1×10^{20} cm^{-3} yielded a resistivity of 1.3×10^{-6} Ω cm^2, the lowest value reported to date for a nonalloyed metal contact directly to GaAs.

The space-charge density required to obtain this resistivity is roughly 10^{20} cm^{-3}, equal to the Si content and roughly ten times the maximum electron density observed with Si doping. Silicon is amphoteric; thus its donor-to-acceptor ratio varies inversely with the square of the electron density. The authors argue that, because midgap Fermi-level pinning occurs at the GaAs surface during MBE growth, the surface depletion region is depleted of electrons, causing the silicon to locate on gallium sublattice sites and behave as donors. Once in the bulk, the Fermi level rises into the conduction band, and some of the silicon atoms switch to arsenic (acceptor) sites, resulting in a maximum electron density around 10^{19} cm^{-3}. These results indicate that surface conditions can alter dopant behavior and that measurements of bulk characteristics can incorrectly predict the surface-related properties that are important to ohmic-contact technology.

VII. Summary

Models relating to the formation of Schottky barriers were examined in terms of effectiveness in explaining experimental barrier heights. All were found to have serious deficiencies. A new model was developed that modified the standard work-function model to explain nearly all experimentally observed barriers to III–V surfaces. A key feature of this model, called the effective-work-function model, is to replace the work function of the applied metal with an appropriately averaged work function of the material which forms the actual interface with the semiconductor surface. In special cases this material will be the same as the applied metal. However, for most cases of technical interest, the dominant material is elemental group-V element As on GaAs, and thus the barrier height is determined by the work function of the common group-V species.

As a result of this understanding, a new ohmic-contact structure for GaAs was developed. A graded bandgap layer of GaInAs is epitaxially grown on GaAs by MBE. This layer is doped *n*-type, and the composition is pure GaAs at the GaAs–graded layer interface and 0–25% InAs at the surface. This structure has no barriers to electron flow, and thus forms a low-resistance ohmic contact with almost any metallurgy without alloying.

Alloyed contacts to *n*-GaAs are simple to fabricate, usable, but far from ideal. This contact process is more "black art" than science. The measured contact resistivity is limited by geometrical effects at the interface, and will probably not be reduced below currently attained values. More refined techniques are becoming available which promise lower resistance and better reliability, but at the cost of greater complexity.

References

1. W. Schottky, *Z. Phys.* **118,** 539 (1942).
2. J. Van Laar, A. Huijser, and T. L. Van Rooy, *J. Vac. Sci. Technol.* **114,** 894 (1977).
3. R. K. Swank, *Phys. Rev.* **153,** 844 (1967).
4. W. E. Spicer, I. Lindau, P. Skeath, and C. Y. Su, *J. Vac. Sci. Technol.* **17,** 1019 (1980).
5. C. A. Mead and W. G. Spitzer, *Phys. Rev.* **34,** A713 (1964).
6. J. Bardeen, *Phys. Rev.* **71,** 717 (1947).
7. S. Kurtin, T. C. McGill, and C. A. Mead, *Phys. Rev. Lett.* **22,** 1433 (1969).
8. J. O. McCaldin, T. C. McGill, and C. A. Mead, *Phys. Rev. Lett.* **36** (1976).
9. S. G. Louie, J. R. Chelikowsky, and M. L. Cohen, *Phys. Rev. B: Solid State* [*3*] **13,** 2461 (1976).
10. J. Tersoff, *Phys. Rev. Lett.* **52,** 465 (1984).
11. L. J. Brillson, *Phys. Rev. Lett.* **40,** 260 (1978).
12. A. Zur, T. C. McGill, and D. L. Smith, *Phys. Rev. B: Condens. Matter* [*3*] **28,** 2060 (1983).
13. J. M. Woodall, C. Lanza, and J. L. Freeouf, *J. Vac. Sci. Technol.* **15,** 1436 (1978).
14. K. Okamoto, C. E. C. Woods, and L. F. Eastman, *Appl. Phys. Lett.* **15,** 636 (1981).
15. K. H. Hsieh, M. Hollis, G. Wicks, C. E. C. Woods, and L. F. Eastman, *Conf. Ser.—Inst. Phys.* **65,** 165 (1983).
16. P. Skeath, C. Y. Su, I. Hino, I. Lindau, and W. E. Spicer, *Appl. Phys. Lett.* **39,** 349 (1981).
17. J. L. Freeouf, J. M. Woodall, and L. M. Foster, *Bull. Am. Phys. Soc.* [*2*] **26,** 285 (1981); J. L. Freeouf and J. M. Woodall, *Appl. Phys. Lett.* **39,** 727 (1981).
18. R. L. Farrow, R. K. Chang, S. Mroczkowski, and F. H. Pollak, *Appl. Phys. Lett.* **31,** 768 (1977).
19. G. P. Schwartz, G. J. Gualtieri, J. E. Griffiths, C. D. Thurmond, and B. Schwartz, *J. Electrochem. Soc.* **127,** 2488 (1980).
20. L. G. Meiners, D. L. Lile, and D. A. Collins, *J. Vac. Sci. Technol.* **16,** 1458 (1979).
21. H. H. Wieder, *J. Vac. Sci. Technol.* **15,** 1498 (1978).
22. A. Hiraki, K. Shuto, S. Kim, W. Kammura, and M. Iwami, *Appl. Phys. Lett.* **31,** 611 (1977).
23. R. S. Williams, private communication.
24. K. W. Frese, Jr., *J. Vac. Sci. Technol.* **16,** 1042 (1979).
25. A. Williams, private communication.
26. J. L. Freeouf, T. N. Jackson, S. E. Laux, and J. M. Woodall, *Appl. Phys. Lett.* **40,** 634 (1982).
27. N. Braslau, *Thin Solid Films* **104,** 391 (1983); *J. Vac. Sci Technol., B* [2] **1,** 700 (1983).
28. N. Braslau, *J. Vac. Sci. Technol.* **19,** 803 (1981).
29. C. Y. Chang, Y. K. Fang, and S. M. Sze, *Solid-State Electron.* **14,** 541 (1971).
30. J. R. Dale and R. G. Turner, *Solid-State Electron.* **14,** 541 (1971).
31. J. B. Gunn, *IBM J. Res. Dev.* **8,** 141 (1964).

32. N. Braslau, J. B. Gunn, and J. L. Staples, *Solid-State Electron.* **10,** 381 (1967).
33. R. H. Cox and H. Strack, *Solid-State Electron.* **10,** 1213 (1967).
34. D. C. Miller, *J. Electrochem. Soc.* **127,** 467 (1980).
35. A. Christou, *Solid-State Electron.* **22,** 141 (1979).
36. B. L. Sharma, *in* "Semiconductors and Semimetals" (R. K. Willardson and A. C. Beer, eds.), Vol. 15, p. 1. Academic Press, New York, 1981.
37. R. P. Gupta and J. Freyer, *Int. J. Electron.* **47,** 459 (1979).
38. S. Tiwari, private communication.
39. W. T. Anderson, A. Christou, and J. E. Davey, *IEEE J. Solid-State Circuits* **SC-13,** 430 (1978).
40. K. Heime, U. König, E. Kohn, and A. Wortmann, *Solid-State Electron.* **17,** 835 (1974).
41. M. Ogawa, *J. Appl. Phys.* **52,** 406 (1980).
42. H. M. Macksey, *Conf. Ser.—Inst. Phys.* **33b,** 245 (1976).
43. M. Yoder, *Solid-State Electron.* **23,** 117 (1980).
44. K. Ohata and M. Ogawa, *Annu. Proc., Reliab. Phys.* [*Symp.*] **12,** 278 (1974).
45. F. Vidimari, *Electron. Lett.* **15,** 675 (1979).
46. M. Heiblum, M. I. Nathan, and C. A. Chang, *Solid-State Electron.* **25,** 185 (1982).
47. N. Yokoyama, S. Ohkawa, and H. Ishikawa, *Jpn. J. Appl. Phys.* **14,** 1071 (1975).
48. G. Eckhardt, *in* "Laser and Electron Beam Processing of Materials" (C. W. White and P. S. Peercy, eds.) p. 467. Academic Press, New York, 1980.
49. G. Badertscher, R. P. Salathe, and W. Luth, *Electron. Lett.* **16,** 113 (1980).
50. H. H. Berger, *J. Electrochem. Soc.* **119,** 509 (1972); *Solid-State Electron.* **15,** 145 (1972).
51. S. J. Procter, L. W. Linholm, and J. A. Mazer, *IEEE Trans. Electron Devices* **ED-30,** 1535 (1983).
52. R. S. Popovic, *Solid-State Electron.* **21,** 1133 (1978).
53. T. J. Lewis, *J. Appl. Phys.* **26,** 1405 (1955).
54. T. S. Kuan, *J. Appl. Phys.* **54,** 6952 (1983).
55. T. J. Magee, J. Peng, J. D. Hong, V. R. Deline, and C. A. Evans, Jr., *Appl. Phys. Lett.* **35,** 615 (1979).
56. S. H. Wemple and W. C. Niehaus, *Conf. Ser.—Inst. Phys.* **33b,** 262 (1977).
57. K. Ohata, T. Nozaki, and N. Kawamure, *IEEE Trans. Electron Devices* **ED-24,** 1129 (1978).
58. W. J. Devlen, C. E. C. Wood, R. Stall, and L. F. Eastman, *Solid-State Electron.* **23,** 823 (1980).
59. J. V. DiLorenzo, W. C. Niehaus, and A. Y. Cho, *J. Appl. Phys.* **50,** 952 (1979).
60. W. T. Tsang, *Appl. Phys. Lett.* **33,** 1022 (1978).
61. P. A. Barnes and A. Y. Cho, *Appl. Phys. Lett.* **33,** 651 (1978).
62. R. L. Mozzi, W. Fabian, and I. J. Piekarski, *Appl. Phys. Lett.* **35,** 337 (1979).
63. Y. I. Nissim, J. F. Gibbons, and R. B. Gold, *IEEE Trans. Electron Devices* **ED-28,** 607 (1981).
64. R. Stall, C. E. C. Wood, K. Board, and L. F. Eastman, *Electron. Lett.* **15,** 800 (1979).
65. J. Massies, J. Chaplart, M. Laviron, and N. T. Linh, *Appl. Phys. Lett.* **38,** 693 (1981).
66. C. A. Mead and W. G. Spitzer, *Phys. Rev. Lett.* **10,** 471 (1963).
67. J. N. Walpole and K. W. Nill, *J. Appl. Phys.* **42,** 5609 (1971).
68. UPS studies of the In 4d core level in our laboratory have verified this interpretation of the transport data.
69. J. M. Woodall, G. D. Pettit, T. N. Jackson, C. Lanza, K. L. Kavanagh, and J. W. Mayer, *Phys. Rev. Lett.* **51,** 1783 (1983).

70. Chin-An Chang, M. Heiblum, R. Ludeke, and M. I. Nathan, *Appl. Phys. Lett.* **39,** 229 (1981).
71. K. Kajiyama, Y. Mizushima, and S. Sakata, *Appl. Phys. Lett.* **23,** 458 (1973).
72. J. M. Woodall, J. L. Freeouf, G. D. Pettit, T. Jackson, and P. D. Kirchner, *J. Vac. Sci. Technol.* **19,** 626 (1981).
73. P. D. Kirchner, *Electron. Mater. Conf., 1984* (1984); P. D. Kirchner, T. N. Jackson, and J. M. Woodall, to be published.

Author Index

Numbers in parentheses are reference numbers and indicate that an author's work is referred to although the name is not cited in the text. Numbers in italics show the page or which the complete reference is listed.

A

B

C

D

E

I

J

K

L

M

N

O

P

R

S

T

Z

Subject Index

B

C

D

E

F

G

H

I

K

L

M

R

S

T